MR. ROLAND M. CRONSHAW
& DR. LYNNE BARRATT
WILTON COTTAGE
FULL SUTTON
YORK YO4 1HW
(0759) 71573

MANGROVE

The Forgotten Habitat

Jeremy Stafford-Deitsch

Mangrove: the Forgotten Habitat is published by Immel Publishing

Copyright: © 1996 text: Jeremy Stafford-Deitsch
Copyright: © 1996 photographs: Jeremy Stafford-Deitsch

The right of Jeremy Stafford-Deitsch to be identified as the author of this work has been asserted in accordance with the English Copyright Design and Patents Act 1988, sections 77 and 78.

Design and typesetting: Jane Stark
Production: Paula Casey-Vine

Cataloguing in Publication Data
A CIP catalogue record for this book is available from the British Library.

ISBN 0 907151 93 0

Immel Publishing Limited
20 Berkeley Street, Berkeley Square, London W1X 5AE
Tel: 0171 491 1799, Fax: 0171 493 5524

To Russell Hanley
who introduced me to
the fascinating world
of mangroves.

*The mangrove or green-backed heron (*Butorides striatus*) is found in mangroves throughout the world, although it is by no means confined to mangal. There are numerous different races and colour variations within the species. This particularly striking bird was photographed in Kilgwyn Swamp, Tobago.*

*A pair of brightly-coloured shield-backed bugs (*Lampromicra senator*) copulate on the branch of a sandy mangrove.*

ontents

The red mangrove trees of Caroni Swamp, Trinidad, at dawn.

Introduction

What would the world be, once bereft
Of wet and of wildness? Let them be left,
O let them be left, wildness and wet;
Long live the weeds and the wilderness yet.
                            - Gerard Manley Hopkins.

When I first considered writing a popular book on mangroves the general reaction from lay people and publishers alike was one of polite disinterest. Mangroves? Mangroves! They are just hot, bug-infested swamps that clutter up the coastline. Who could possibly find such unpleasant places interesting? What is there to see apart from mud, funny-looking trees and endless biting insects?

In many tropical coastal areas where tourists go sightseeing they are shown the coral reefs and beautiful beaches but the mangroves – if they have not already been cleared 'to improve the view' or to provide more land for building

– are actively avoided. This outlook still predominates despite the fact that mangrove forest explorers know this strange habitat is rich in quiet secrets.

For Europeans in the past mangroves encapsulated the very worst features of the tropics: blinding heat, dangerous creatures and a virtually impenetrable barrier to the lands they had travelled so far to exploit and plunder. This attitude persists and is all too easily reinforced by those who wish to 'improve' coastlines by destroying the mangroves. But it is not the only attitude: the indigenous peoples of the tropics have long appreciated the host of foods, medicines and quality timber that the mangroves contain. Where their opinions have not been overpowered by the pressures of 'development', mangroves still survive.

A great many people helped in many ways in turning the idea of this book into actuality. Numerous members of the scientific staff at the Northern Territory Museum of Arts and Sciences in Darwin, Australia, provided crucial assistance on many occasions, and without that help the project would have been still-born. First and foremost is Russell Hanley who spent a great many hours taking me into the mangroves and indefatigably pointing out and re-explaining features of mangrove biology that I consistently failed to comprehend. Other helpers in the field or in the vital task of identifying mangrove beasts in the Northern Territory include Gabi Caswell, Adèle Williams, Daniel Couriel, Helen Larson, Richard Willen, Graham Brown, Sandy Bruce, Robert McMahon and the members of the Marine Ecology Unit of the museum. Others whose help was invaluable include Kay and Rodney Fox and Greg Harman in Australia, Philip and Katherine Munday, Max Benjamin and Robert Halstead in Papua New Guinea, as well as Winston Nanan in Trinidad, Samuel Gruber and Rocky Strong in the Bahamas, Jack Randall, Jim Breakell and the entomolgists of the British

Museum of Natural History. Last but by no means least I would like to thank Peter Vine (himself a marine biologist) of Immel Publishing whose enthusiasm for this project only increased as it neared completion.

As I am not by training a scientist I have relied heavily on numerous texts for information. These are listed in the bibliography. I would like to single out for special praise P. B. Tomlinson's *The Botany of Mangroves*. I recommend this grand work to anyone interested in studying the plants of the mangrove forest in greater detail. For those wishing to know more about the scale of the problem facing mangrove forests I recommend M. G. Wenban-Smith's *Global Status and Extent of Mangrove Forests*, a report prepared for the United Kingdom branch of the World Wide Fund for Nature.

It is my earnest hope that this volume will contribute to the growing groundswell of public outrage at the headlong destruction of the world's mangrove forests. If it shows the doubters that mangroves are inherently fascinating and biologically invaluable then it will achieve more than I could have hoped for.

The uniqueness of the mangrove habitat can only be explained in fairly technical terms. It is one thing to say that the trees are adapted to the intertidal habitat and that they can survive regular periods of salt water immersion. But if one wants to know more than this then the discussion inevitably becomes more technical. What are the specific problems facing these trees? How do the different species adapt to the challenges? Although this book is aimed at as wide a range of readers as possible, it is important to cover such technical topics in the depth appropriate for explaining the adaptations of mangroves since it is these adaptations that make them unique and of such special importance. In fact the principles involved are rarely difficult to grasp. Hopefully this book will crystallise and make accessible the theories that are at present

*Black-bellied whistling ducks (*Dendrocygna autumnalis*) are permanent residents of Buccoo Swamp in Tobago where they feed on the vegetation.*

locked away in specialist scientific publications, without over-simplifying them. Anyone wishing to study the subject in further detail can consult the bibliography.

Having acknowledged my debt to the scientific community I should of course add that any errors of fact contained within these pages are entirely their own.

Exploring the Mangrove World

What is it like to explore the mangrove habitat and how is the exploration best achieved? This is not an environment to be explored lightly, but a few simple guidelines can make the undertaking fascinating, enjoyable and safe. A mangrove swamp is suffused with an almost alien aura thanks to the odd-looking trees, the coming and going of the sea and the vast expanse of mud. I remember a review of a book on modern physics in which the reviewer said that he more or less recognised the physics as it was discussed in the book but it was not quite right. He said it was rather like walking down a street he was familiar with and seeing that everything had been painted purple. This is exactly the surreal effect of entering a mangrove swamp. Yes, there are trees everywhere but such funny-looking trees. They are the trees of a forest one might enter in one's dreams, a forest in a parallel universe, in which everything has evolved more or less the same – the trees have trunks, branches and green leaves – but there are a number of odd-looking features about them. And in a

View through stilt-rooted mangroves in the Northern Territory, Australia.

A mangrove-lined creek of the Adelaide River in the Northern Territory.

sense they *are* alien. To all physiological intents and purposes they do inhabit another world: the intertidal world.

And then there is the silence. It is the silence an intruder experiences, the silence a burglar must feel when he breaks into a home. But in the mangroves it has something ancient, something other, something separate about it. One has fought one's way into this world and its numinosity cocoons one's anthropocentrism. It is akin to the silence of libraries: to whisper to a companion is acceptable, to talk improper, to shout a crime. It is not, to be sure, a *silent* silence; it is punctuated with activity. There are the pistol-shots of strange shrimps snapping their giant claw. There are the constant muffled slushings and crunchings of life in the soft mud. And there is the endless chirpy babble of what the biologists of the Northern Territory call snibs: small, nearly invisible birds.

Sometimes one hears a noise one cannot identify and the imagination runs riot. I was once sure I had heard a pterodactyl woosh through the canopy of a Papuan mangrove swamp. I could not see the creature of course, but the setting seemed to have a prehistoric solemnity about it. (I was later informed that it must have been a hornbill but when it comes to cryptozoology I have always been something of a soft touch!)

In some mangrove swamps there *are* animals older than the dinosaurs which it is advisable to avoid. In the creeks and rivers of Australia's Northern Territory mangroves two extremely dangerous animals are all too common: the estuarine crocodile (also known as the salt water crocodile) and the box jellyfish. The danger posed by the former (the largest extant crocodile) hardly needs explaining. The latter is a type of jellyfish (although not a member of the class of 'typical' jellyfishes) with stinging tentacles so powerful that it can quickly kill a human. For just these two reasons attempting to swim in a Northern Territory mangal would be flirting with suicide, so clearly anyone wishing to explore a mangal should have local knowledge of any dangers.

The estuarine crocodile is also found in Papua New Guinea and recently I had the opportunity to snorkel in various mangroves there. I made myself unpopular by constantly asking if there were any crocodiles in the area. The water was clear enough to take pictures – though one had to move extremely carefully to avoid silting everything up. This gave me the opportunity to photograph the

A large estuarine crocodile (Crocodylus porosus) resting on the exposed mud flats of the Mary River, Northern Territory. Saltwater incursion has made this stretch of river available for colonisation by mangroves: note the grey mangrove saplings growing in the mud.

marine creatures of a Pacific mangrove, something I could not do in the Northern Territory. On one occasion I was snorkelling along the edge of a remote patch of mangroves accompanied by Ron Taylor the celebrated Australian underwater film-maker. We were swimming towards a little mangrove-fringed inlet, and the visibility was getting worse and worse; the underwater scenery of mangrove prop roots and sea grass looked particularly sinister. My imagination began to run wild, but fortunately we did not come across any crocodiles.

An underwater view in unusually clear water of the prop roots of mangroves bordering an area of sea grass in Papua New Guinea. The proximity of different ecosystems can lead to a considerable overlap of animal species between the two.

The danger posed by estuarine crocodiles is very real and they can be found almost anywhere in their range, from low-lying fresh-water locations to the open ocean: Philip Munday, a diver and marine biologist working in New Britain, Papua New Guinea, encountered one swimming in open water off a river mouth in Kimbe Bay. He and I spent some time exploring the Kimbe Bay mangroves underwater, and shortly after I left he sent me a newspaper clipping telling the tragic story of a fourteen-year-old girl who had been killed by an estuarine crocodile while on a shellfish-gathering expedition in some mangroves just a little further down the coast. Because they move around a lot, as dominant males constantly evict weaker or smaller males from prime territories, estuarine crocodiles can turn up anywhere, and even local knowledge may not be adequate.

Another problem when exploring a mangrove creek in a small boat is that it is easy to get stuck on a mudbank. This is especially likely if the tide is going out. One then has to sit and wait several hours for the incoming tide. Once I was exploring the Mary River, Northern Territory, mangroves in a small boat with Russell Hanley, a marine biologist from the Northern Territory Museum. We approached one mudbank that was absolutely jumping with mudskippers and fiddler crabs. I was keen to try to photograph them, but of course as soon as I came within photographic range everything dived into its hiding place in the mud and we had to wait for them to reappear. We waited only a few minutes and then realising that the tide was going out, we decided to leave. The motor roared into life but the boat did not budge – we had already settled onto the mud.

The idea of spending many hours trapped in a tiny boat on a mudbank on a crocodile-infested river did not appeal to either of us. With every second the boat settled more firmly into the mud. Getting out of the boat and trying to push it from within the mud was hopeless – and frightening given the presence of crocodiles. They slip quietly into the water when they see people coming and vanish from sight. The only thing more frightening than a visible crocodile is an invisible one. Somehow we managed to get the boat off the mud and proceeded on our way, but it was an educational experience; just a few more minutes and we would have been stranded. The next day when Russell and I were in the boat looking for a place to enter the mangroves we picked a spot and gingerly approached. The old 'moving log' story is all too true. A huge 'log' in the mangrove roots suddenly moved and slithered across the mud into the river. So we picked another spot to enter.

Biting insects can also make a visit to the mangroves anything from unpleasant to dangerous. On one occasion I wanted to photograph the mudskippers on a mudflat in Darwin Harbour, a huge, creek-fringed, mangrove-bordered inlet over 10 km wide at its widest point. I was dropped off at a place that I had visited several times before. Previously I had not been much bothered by insects. I struggled out on to the mud, set up my camera and began the long wait for the creatures of the mud to emerge. I noted that the atmosphere seemed to be taking on a grey tinge; perhaps a cloud was passing in front of the sun. I glanced at the sun. There was no cloud. Then I began to feel innumerable tiny, itchy, burning pinpricks all over my exposed skin (except, strangely enough, my face). A close examination revealed that it was covered in biting midges. I realised that I had not brought my insect repellent with me and tried slapping at them, as well as smothering my skin in mud. But I was finally forced to retreat to the shore. Every square millimetre of exposed skin had been bitten.

(I later found that my insect repellent had been in my back pocket all along!)

Biting midges and mosquitoes are common in mangroves, and mosquitoes are known to feed ravenously on the hapless mudskippers. The midges deposit their eggs in the mud, while mosquitoes lay theirs in fresh or brackish water; a favourite place for mosquito larvae to develop is in the hollows of rotting logs. In some parts of the world the mosquitoes in the mangal carry malaria or other diseases. Protective clothing and insect repellent are therefore vital.

There are two basic methods of entering a mangrove forest. The first is from the seaward edge in a small boat and the second is from the landward edge on foot. Both are best done when the tide is out. Given the density of floral obstacles in the mangal and the soft mud that is usually underfoot, a good knowledge of the rate at which the tide is coming in is essential – because of the slowness of one's progress, one can easily become cut off from one's route of exit. Creek banks that are made of soft mud are hard enough to cross at low tide, and they fill rapidly as the tide rises so that one can find oneself doomed to the indignity of climbing a tree and waiting for the tide to go out again.

Because many species of mangrove plant form monospecific stands – equally sized, equally spaced, indistinguishable trees – it is all too easy to become completely lost, or to lose anything left a short distance away. On one occasion in Trinidad's Caroni Swamp my guide had taken me deep into the bowels of the swamp to where egrets and herons were nesting. I wanted to photograph some chicks in their nests so I waded with various cameras towards the cacophany of the nesting birds. When I had established which camera would be best, I left the other at the base of a tree. I knew the tide was going out: we had seen it doing so on our way in. When I had taken the shots of the chicks in their nests I headed back to where I had left my camera. But the trees all looked the same. Furthermore, if the tide really was going out,

Opposite: Red mangrove aerial roots form an almost impenetrable barrier in Trinidad's Caroni Swamp. Given the typical substrate of a mangrove swamp - soft mud - and the closely-packed tree trunks and root systems, progress through a mangrove swamp is slow and exhausting work.

Entering a mangrove swamp of Australia's Northern Territory from the landward edge, one is likely to encounter an area densely packed with indistinguishable yellow mangrove trees. Precautions should be taken against getting lost in this zone of seemingly identical trees.

why was there no wet mark above the waterline on the trees? I realised with more than a touch of panic that although the tide may have been going out in the main channels of the swamp, here it was still coming in. And the reason I could not see my camera and its priceless telephoto lens was because... I swallowed hard. The possibility did not bear thinking about. I searched and searched until I found them – centimetres from the rising tide!

Getting lost in the mangroves is all too easy. One can follow one's footprints if the substrate is soft mud, but once one takes a wrong turn one is just as likely to end up following one's *lost* footprints. And if the tide is coming in, footprints are soon submerged. As a basic starting point it is a good idea to note the position of the sun relative to one's progress through the mangroves. Another, better solution is to enter this floral labyrinth armed Theseus-like with ribbon to mark the route through the maze. It is important, however, to remove the ribbons once you have finished your exploration. It is surprising how many well-used mangrove forests are still marked with tape and ribbon long after the user has departed.

Another important instrument is a compass. No matter how lost one is, a compass will show the general direction of the sea or the land. If you are exploring the creeks of a mangrove swamp in a small boat always try to get a local guide to take you. What looks unrecognisable to you will be familiar to your guide.

Insect repellent and sunscreen are obligatory. Moreover, because of the often stultifying heat that can build within a mangal and the sheer physical exertion required to progress through it, a good supply of water must be carried and a wide-brimmed hat should be worn. A light pair of binoculars is also recommended for observing distant, timid animals.

Footwear should be strong-soled and capable of withstanding being water-logged and smothered in mud. The best solution is a pair of thick-soled wetsuit

*The soft mud of the mangrove swamp is teeming with creatures that can be crushed by walking through it. Australia's blue nipper crab (*Thalamita crenata*) is an aggressive mud-dweller that can retaliate with greater ferocity than the term 'nipper' might suggest.*

bootees. Anything bulkier, such as boots, will become too heavy because too much mud will cling to them.

Given the soft, creature-infested mud and the fragile root systems of the trees, there is a very real problem of damage by people moving through the mangroves. A regular procession of feet trampling over the semi-submerged roots of a tree (often the easiest way to make progress) could kill it.

In various parts of the world walkways are being constructed through the mangroves. These are plank-covered paths above the tidal height, which allow people to walk in comfort and safety above the mud. They are to be encouraged as they do little damage and allow visitors to see the mangroves at first hand without difficulty to themselves or damage to the habitat. Certainly they are the only practical way to introduce children – the next generation of biologists and conservationists – to this otherwise difficult habitat and so convince them of the mangrove ecosystem's intrinsic importance both for study and conservation. The *tabula rasa* of a child's mind is indefatigably eager to embrace every corner of the natural world and children are the mangrove ecosystem's hope for the future.

The Plants
of the Mangrove Forest

Defining the habitat

Fundamental to an understanding of the mangrove habitat is the significance of the tide. As we will see later, many of the plants found in the mangrove swamp can be found in habitats unaffected by the tide, but the mangrove swamp itself is a tropical or subtropical forest that occurs on low-lying ground which is subjected to tidal immersion. In other words the mangrove forest is an intertidal forest. At low tide the substrate (usually mud) into which the trees are anchored is exposed, at high tide it is covered by sea water. Variations in the height of the tide will be reflected in the amount and regularity of substrate submergence and exposure. But while the entire mangrove ecosystem is within the range of the tide it is not limited to that part of the shoreline submerged and exposed by every tide. On the contrary, its penetration inland is limited only by the height of the *highest* tide.

The term 'mangrove' is often applied to the various plants that occur

Top: A tidal creek in Darwin Harbour at high tide. Bottom: The same creek at low tide. Four different species of mangrove are visible. In the left foreground the low, shrub-like plants are the club mangrove. Immediately behind them and taller is a young mangrove apple tree. The larger tree in the upper right of the picture is a grey mangrove. On the opposite bank is a line of stilt-rooted mangroves. The three species of mangroves in the foreground are typical of the seaward edge. The stilt-rooted mangrove is commonly found lining creek banks.

within the intertidal range. However one must distinguish between those plants that rarely if ever occur beyond this range and those that can be found both within the mangrove swamp and outside it. Salt water immersion kills the vast majority of terrestrial plants, but different species can survive varying amounts of salt water, both in terms of the time submerged and in terms of the concentration of the salt in the water. Tropical river systems often have luxuriant vegetation on their banks and in their floodplains. Varying brackish conditions will occur in the intertidal areas and this can be reflected in vegetation that cannot survive submergence in pure sea water. Conversely mangroves can penetrate inland along river banks tidally subjected to rising and falling water levels but sufficiently far from the sea for there to be little if any dissolved salt in the water.

The mottled green and brown bark and the winding ribbon-like roots of this tree in Papua New Guinea identify it as a cannonball mangrove. This species cannot tolerate dissolved salt of sea-water strength and is found at the landward edge of mangrove forests subjected to a regular supply of fresh water.

There are at least two uses for the term 'mangrove': the intertidal forest ecosystem and the plant constituents of the ecosystem. A term often used to refer to the ecosystem itself is mangal. At first glance this sounds circular: the plants define the ecosystem and the ecosystem defines the plants. However, as I have said, one can distinguish between those plants that are virtually confined to the mangal (true mangrove plants) and those that occur both there and elsewhere (mangrove associates).

Unfortunately such a definition is not perfect for two reasons. First, the 'true' mangrove plants need not (and tend not to) occur in all areas of the mangal, and secondly some of the true mangrove plants occasionally occur both above and below the reaches of all tides. The reasons for the second

*Mangrove forests halt at the height of the highest tide. Here, on the banks of the Adelaide River in the Northern Territory, there is an abrupt transition from mangroves at the left to grass and finally to terrestrial paperbark forest (*Malaleuca leucodendron*). The mangroves are typical landward-edge species: the sandy mangrove and yellow mangrove.*

situation will become clear but the main point is that these plants have specific adaptions to the intertidal habitat (which we will consider later), and these limit the success if not the extent of their range.

Another way of attempting to define the mangal is in terms of the effect of the mangrove plants on the environment in which they become established, in other words the extent to which true mangrove plants actually establish conditions in which other species can become established. This proposal hinges, as we shall see later, on certain controversial theories about the relationship between true mangrove plants and land formation.

The difficulty of determining what actually *is* a mangrove plant as opposed to a mangrove associate has led to considerable variation in the numbers of plants classified as true mangrove plants. One account lists 400 arborescent and 500 non-arborescent plants in the Indo-West Pacific region alone, while another lists 94 species of mangrove plant and mangrove associate worldwide. Many types of plant are found in the mangal. In the mangrove swamps of the Northern Territory, Australia, the following are common constituents: evergreen and deciduous trees, perennial and evergreen shrubs, epiphytes, parasites and climbers, perennial forbs, perennial grasses, palms and perennial ferns.

The small-leafed mangrove is better classified as a mangrove associate: it can occur at the back edge of the mangal where it can form extensive thickets as it does here in the Northern Territory, or it can occur in sand dune or rocky areas above the height of the tide.

Tomlinson has listed a number of criteria for distinguishing a true mangrove from a mangrove associate. The true mangrove plants are identified 'because they possess all or most of the following features:

1. Complete *fidelity* to the mangrove environment: that is, they occur only in mangal and do not extend into terrestrial communities.
2. A *major role* in the structure of the community and the ability to form pure stands.
3. *Morphological specialization* that adapts them to their environment: the most obvious are *aerial roots*, associated with gas exchange, and *vivipary* of the embryo, whose functional significance is not clear.

Epiphytes are common in mangrove forests. This example (Cynanchum carnosum) *is from the Northern Territory.*

4. Some physiological mechanism for *salt exclusion* so that they can grow in sea water; they frequently visibly excrete salt.

5. *Taxonomic isolation* from terrestrial relatives. Strict mangroves are separated from their relatives at least at the generic level and often at the subfamily or family level. For minor mangroves, the isolation is mostly at the generic level.'

The notion of fidelity is self-explanatory. However it is not true that true mangroves can *only* survive in the intertidal habitat. Rather, that is where they are almost exclusively found. Their adaptations to this harsh environment (especially salt water) place them at a competitive disadvantage in other, less harsh, terrestrial habitats. Mangroves grow perfectly well with only fresh water bathing them (as botanical experiments have shown), but they are usually outdone by non-mangrove plants in non-intertidal habitats. They also lack mechanisms for becoming established in such habitats. They are therefore called facultative as opposed to obligate halophytes (plants that lives in saline conditions). This means that they do not *have* to live where they do, but they are adapted to do so to an extent that does not allow them to compete with non-mangrove plants in other habitats.

The requirement that they play a major role in the habitat and have the ability to form pure stands is immediately apparent to anyone who ventures into a healthy mangrove swamp. The uniformity of the trees is often a striking feature. If one tries to enter a mangal in the Northern Territory of Australia from the salt pan area at its back one is likely to encounter a dense wall of the yellow mangrove (*Ceriops tagal*). The seaward edge may be an elegantly spaced

The prop-rooted mangroves are archetypal members of the mangrove community fulfilling Tomlinson's criteria: they tend only to be found in the mangal, they form pure stands, they possess aerial roots and viviparous seedlings, they are highly efficient at limiting the amount of salt that enters their root systems and they have no close non-mangrove relatives. These plants are stilt-rooted mangroves in typical formation lining a creek bank in the Northern Territory.

An elegant 'parkland' of well-spaced mangrove apples established along the seaward edge of a Northern Territory mangal.

'parkland' of the mangrove apple (*Sonneratia alba*). A creek cutting through the mangroves will, in all likelihood, be lined by the stilt-rooted mangrove (*Rhizophora stylosa*). Apart from a few other mangrove species lining its landward fringes, the Caroni Swamp of Trinidad is a vast, towering and magnificent forest of the red mangrove (*Rhizophora mangle*).

Tomlinson's third point deals with morphological specialisations. Some of these are very conspicuous. The mud in which the trees are anchored is usually anaerobic (lacking oxygen). Many mangroves have conspicuous adaptions that allow them to aerate their roots and thus take in the oxygen they require. The prop-rooted mangroves (the *Rhizophora* genus) have bizarre prop roots given off at all angles from the trunk that form in effect flying buttresses and fulfil a supportive as well as a gas-exchange role. Many species of mangrove have extensions from their underground roots that break through the mud and are thus exposed to the air when the tide is out. Such structures are known as pneumatophores. In the mangrove apple the pneumatophores can be thicker than a man's arm and over two metres in height. In the grey mangrove (*Avicennia marina*) the pneumatophores are considerably thinner and usually shorter. Other mangroves produce what are called knee roots. These look like the knobbly knees of some unfortunate drowned in the mud. The large-leafed orange mangrove (*Bruguiera gymnorrhiza*) produces such root excrescences. Another variation on the theme is provided by the cannonball mangrove (*Xylocarpus granatum*): the upper part of the roots are exposed above the mud and dorso-ventrally flattened into long, meandering ribbon-like structures. So characteristic are the aerial root adaptions of different species of mangrove that they are an easy guide to identification, certainly to the generic level and sometimes even to the species level.

An impressive expanse of knee-rooted mangrove aerial roots. Erosion of the substrate has revealed more of the root structures than is typically visible.

Vivipary, which also occurs in Tomlinson's third requirement, means the development of the embryo while it is still attached to the parent plant. This is highly unusual in seed plants; normally there is a delay between fertilisation and development during which the seed is separated and dispersed from the parent

Left: A developing hypocotyl on a yellow-flowered orange mangrove. The recurved flower structure is thought to discourage insects and attract birds as the pollinating agent.

Right: Virtually fully formed hypocotyls hang from a yellow mangrove tree.

plant. The seedling subsequently develops from the seed. In the case of mangroves, varying degrees of vivipary exist but it is at its most advanced amongst the prop-rooted mangroves and their relatives in the family Rhizophoraceae (the other genera are *Bruguiera, Ceriops* and *Kandelia*). In this group the propagule (the unit of propagation) is not the seed. Instead, the seed develops into the seedling while still attached to the parent tree: the developing seedling stem (called the hypocotyl) grows through the seed coat and out of the fruit: in other words the seedling will become the unit of propagation where in typical seed plants it would be the seed. The dangling hypocotyls developing on a tree of this group are an unforgettable sight.

The question that obviously arises is why so unusual a method of development should have arisen in plants that occupy so difficult a habitat? There is a powerful suspicion that the two are interrelated. Scientists have discovered that the developing embryo of the red mangrove is dependent on its parent for photosynthetic products (starch). However this in itself does not seem sufficient to explain vivipary. More subtle adaptions have been discovered. It has been demonstrated that chloride salts (such as the sodium chloride of sea water) are inhibitors of germination. Thus vivipary may have developed as a means of avoiding exposure by the embryo to the harsh salt-water environment. It may be that there are gland cells between the fruit wall and seed coat of the viviparous prop-rooted mangroves that stop damaging ions from reaching the developing embryo from the parent plant.

Other mangroves demonstrate what has been termed cryptovivipary: while the developing embryo erupts from the wall of the seed it is jettisoned from the parent plant before it breaks through the fruit. Cryptovivipary is demonstrated by, amongst others, the river mangrove (*Aegiceras corniculatum*). Investigation of the connection between the embryo and the mother plant by researchers has

Table 1.

Species list, family and common name of common mangroves and mangrove associates

Species	Common name	Family	Mangrove/Mangrove Associate
Acanthus iliciofolius	Holly-leaf mangrove	Acanthaceae	M A
Acrostichum speciosum	Mangrove fern	Pteridaceae	M A
Aegialitis annulata	Club mangrove	Plumbaginaceae	M
Aegiceras corniculatum	River mangrove	Myrsinaceae	M
Avicennia germinans	Black mangrove	Avicenniaceae	M
Avicennia marina	Grey mangrove	Avicenniaceae	M
Barringtonia racemosa	Freshwater mangrove	Lecythidaceae	M A
Bruguiera exaristata	Yellow-flowered orange mangrove	Rhizophoraceae	M
Bruguiera gymnorrhiza	Large-leafed orange mangrove	Rhizophoraceae	M
Bruguiera parviflora	Small-leafed orange mangrove	Rhizophoraceae	M
Camptostemon schultzii	Schultz's mangrove	Bombacaceae	M
Ceriops tagal	Yellow mangrove	Rhizophoraceae	M
Conocarpus erectus	Buttonwood	Combretaceae	M A
Excoecaria agallocha	Blind-your-eye mangrove	Euphorbiaceae	M

Species	Common Name	Family	Mangrove/Mangrove Associate
Excoecaria ovalis	Oval-leafed blind-your-eye mangrove	Euphorbiaceae	M
Hibiscus tiliaceus	Native hibiscus	Malvaceae	M A
Laguncularia racemosa	White mangrove	Combretaceae	M
Lumnitzera racemosa	Sandy mangrove	Combretaceae	M
Nypa fruticans	Mangrove palm	Arecaceae	M
Osbornia octodonta	Myrtle mangrove	Myrtaceae	M A
Pemphis acidula	Small-leafed mangrove	Lythraceae	M A
Rhizophora mangle	Red mangrove	Rhizophoraceae	M
Rhizophora stylosa	Stilt-rooted mangrove	Rhizophoraceae	M
Sesuvium portulacastrum	Seaside purslane	Aizoaceae	M A
Sonneratia alba	Mangrove apple	Sonneratiaceae	M
Sonneratia lanceolata	Brackish mangrove	Sonneratiaceae	M
Xylocarpus granatum	Cannonball mangrove	Meliaceae	M
Xylocarpus mekongensis	Cedar mangrove	Meliaceae	M

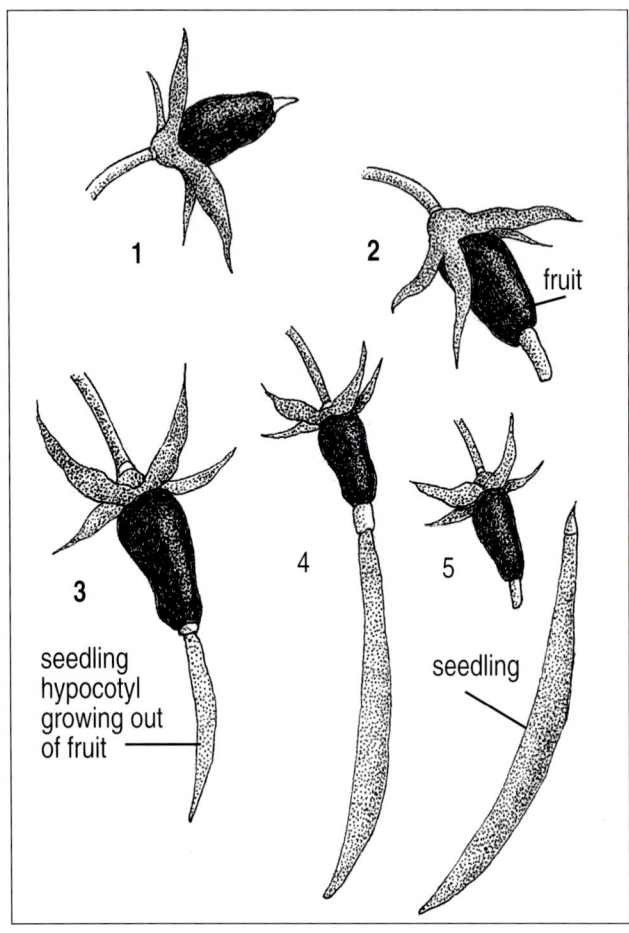

Fig. 1. Development of viviparous seedling.

Table 2: *Common names of mangrove genera.*	
Genus	**Common Name**
Avicennia	Pencil-rooted mangroves
Bruguiera	Knee-rooted mangroves
Excoecaria	Milky mangroves
Lumnitzera	Lumnitzer's mangroves
Rhizophora	Prop-rooted mangroves
Sonneratia	Peg-rooted mangroves
Xylocarpus	Woody-fruited mangroves

supported the theory that vivipary and cryptovivipary have developed as means of overcoming the toxic effects of immersion in sea water. It has been shown that the developing cryptoviviparous embryo is nutritionally dependent on the mother plant for photosynthetic by-products. However, what is more interesting is that there appear to be two glandular tissue walls – the first between the stalk and calyx on the one side and the fruit on the other, the second between the seed coat and the embryo – which act as barriers to the transport of chloride into the embryo. It appears that the barriers gradually allow more chloride ions to enter the embryo as it develops. In effect it is gradually being conditioned to tolerate the chloride it will encounter in sea water when detached from the mother tree.

Item 4 on the list is the requirement for a salt-exclusion mechanism. This typically takes the form of what are known as salt glands on the leaves. These are tiny pores enclosed by specialised cells. Species of mangrove that possess salt glands include the pencil-rooted mangroves (the *Avicennia* genus), the club mangrove (*Aegialitis annulata*), the river mangrove (*Aegiceras corniculatum*) and the holly-leaf mangrove (*Acanthus iliciofolius*). The excreted salt is often visible on the surface of the leaves of such plants.

Salt crystals have formed on the leaf of this club mangrove. This species has salt glands in the leaves that excrete excess salt.

Finally, item 5 of Tomlinson's list, taxonomic isolation, is the claim that the higher the fidelity of a plant to the intertidal habitat the greater the morphological differences between it and its non-mangrove relatives.

The most important criterion for identifying a plant as a true mangrove is probably the first: fidelity to the environment. Other plants can possess other features on the list: for example, salt excretion glands are found in terrestrial halophytes, and aerial roots occur in swamp plants. Mangrove associates can even form pure stands in the mangrove under certain conditions: for example the holly-leaf mangrove and the mangrove fern (*Acrostichum speciosum*) can dominate disturbed areas.

If it has proved more difficult to define the mangal and its plants than we might have expected, nevertheless the two basic features – tidal range and a concentration of certain plants that are almost exclusively confined to the inhospitable ecotone between sea and land – will suffice as a starting point. There are numerous factors such as air temperature, rainfall patterns and soil composition that have their effects on the composition of the mangal. The ecosystem is governed by an interplay of complex processes beyond the scope of a simple definition. It should be mentioned in passing that the term 'freshwater mangrove' is sometimes used to describe certain fresh water plants (including *Barringtonia racemosa*). The term has little meaning if one holds that fidelity to a habitat regularly immersed in sea water is a fundamental characteristic of true mangrove plants. However there are plants found in the mangal that clearly require the influence of fresh water. These occur in the upper reaches of the mangrove swamp or on the banks of river systems where the tide reaches but the salt water does not. As such the term is excusable. An example of another mangrove plant that has very limited tolerance to salt water would be the cannonball mangrove.

The modern distribution of mangals and mangrove species

Mangrove swamps occur at their most prolific between the latitudes of 25° N and 25° S. They do extend some 7° further north in Japan and some 15° further south in New Zealand, Australia and Africa. There are two areas of floral diversity which have been classified as the eastern and western hemisphere regions.

The former ranges through the Indian Ocean to the western Pacific and includes the coasts of East Africa, India, Asia, Indonesia, Malaysia, Australia, Papua New Guinea and the islands of the central Pacific. The greatest number of mangrove species (about 30) is to be found on the tropical Queensland coast of Australia. As one moves eastwards or westwards, so the number of species declines. The total number of true mangroves in the eastern hemisphere region is about 40.

The western hemisphere region basically encompasses the Atlantic Ocean: from the west coast of Africa across to the Caribbean and both the Atlantic and Pacific coasts of North and South America. The total number of true mangroves

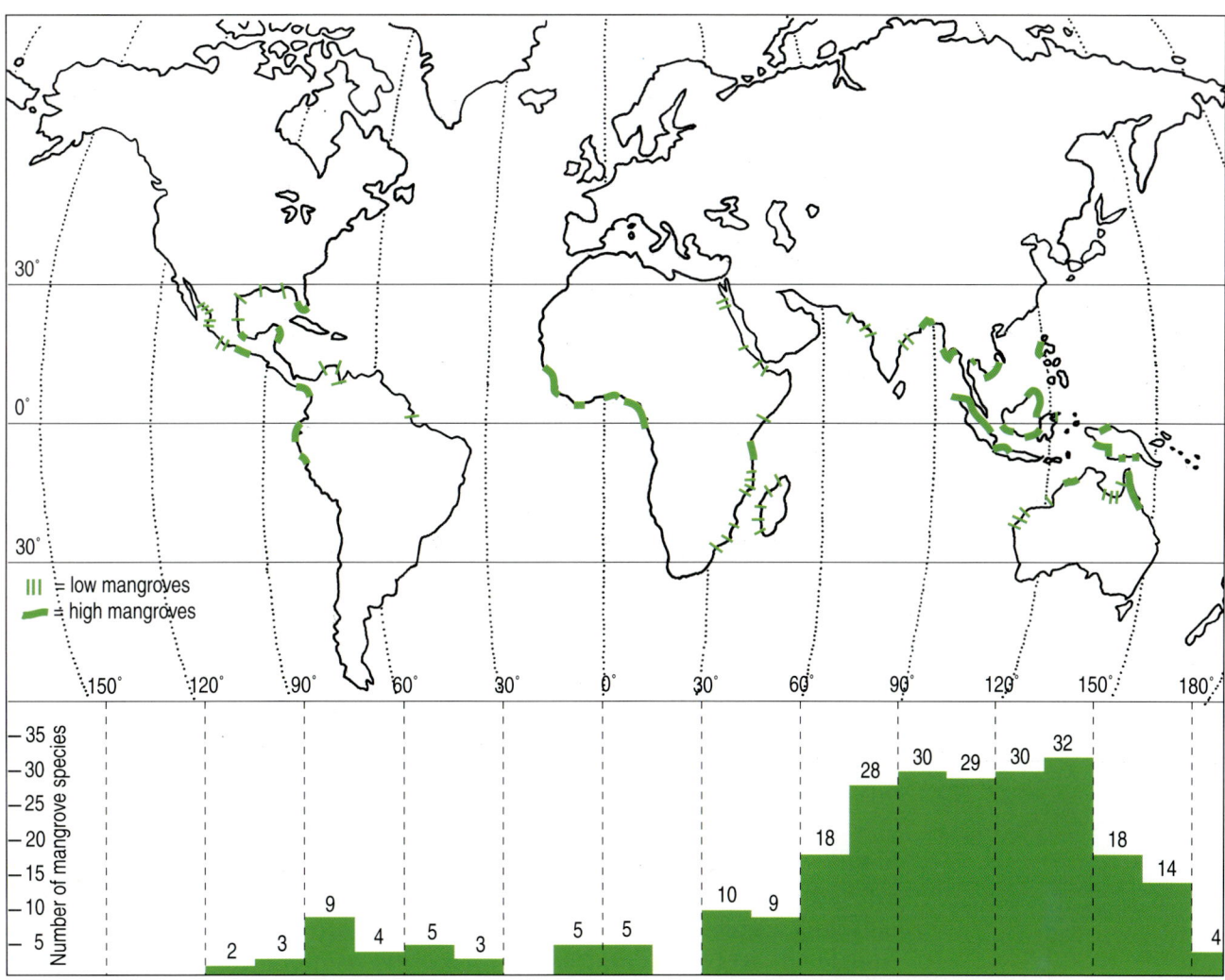

Fig. 2. (After Tomlinson and Blasco combined) World map of mangrove distribution showing species number.

in this area is eight, and the apex of speciation is in the Caribbean.

There is some doubt as to whether any mangrove species is found in both the eastern and western regions. The red mangrove of the western hemisphere may extend into the western Pacific regions but whether the almost identical *Rhizophora samoensis* is in fact the same species is yet to be established. Nevertheless no mangrove species has a worldwide distribution. The large-leafed orange mangrove is one of the most widely distributed of the true mangroves and reaches from the Indian Ocean coast of Africa to Samoa. While no true mangrove has a truly pantropical distribution, some of the plants found in neighbouring coastal habitats (as well as in mangal) do. An example is the seaside purslane (*Sesuvium portulacastrum*). This herb is typically found on sand dunes, beaches and mudflats but it also occurs on the landward edge of the mangal, where it is therefore classified as a mangrove associate.

The considerably greater number of species found in the eastern hemisphere is mirrored in the level of speciation found in neighbouring shallow marine habitats such as coral reefs and sea grass beds - the number of species in these neighbouring habitats is considerably greater in the Indo-Pacific region. Various theories have been put forward to explain the relative paucity in the New World region. It is generally thought that the apex of species in the eastern hemisphere corresponds to the area of a Cretaceous sea, the Tethys Sea, where the ancient ancestors of the modern species originated. The present centres of speciation are thus places where evolution has been going on

The delicate flower of the seaside purslane. While no true mangrove plant has a pantropical distribution, the seaside purslane is an example of a mangrove associate found in tropical coastal habitats worldwide.

longest; the species in the central location are the most recently evolved, while older species have migrated further away from the centre over geological time.

One interesting feature of the mangroves of the eastern region is that the apex of speciation (northern Queensland, Australia) does not correspond with that for coral reefs and sea grasses; the latter is further north, in the waters bordered by the Philippines, Malaysia and Indonesia. Various hypotheses have been suggested to account for this anomaly. Perhaps the Queensland coast mangroves are descendants of ancient species that populated the area when Australia had a more southerly position. The recent land bridge to New Guinea would have given a route for replenishment by newer species from the north, thus providing for both old and new species to co-exist in one habitat.

However if there are problems in accounting for the (relatively short) distances mangroves would have had to travel to establish themselves in northern Australia, they are as nothing compared to the difficulties in explaining how the mangroves of the New World became established and why, furthermore, there are so few species. Given the vast distances involved it is hard to accept that western hemisphere mangroves are connected with those of the eastern hemisphere via a trans-Pacific migration. Alternative explanations have centred on the theory of plate tectonics; or the paucity of species in the New World may have been due to a more limited variety of habitats for mangrove adaptation and evolution.

The factors affecting mangrove distribution

Air temperature is probably the most important factor limiting the spread of mangroves away from tropical latitudes. If one examines the number of mangrove species on the eastern coast of Australia one notices a continuous decline in the number of species the further south one goes.

The highest number of species is found on the northeastern coast between

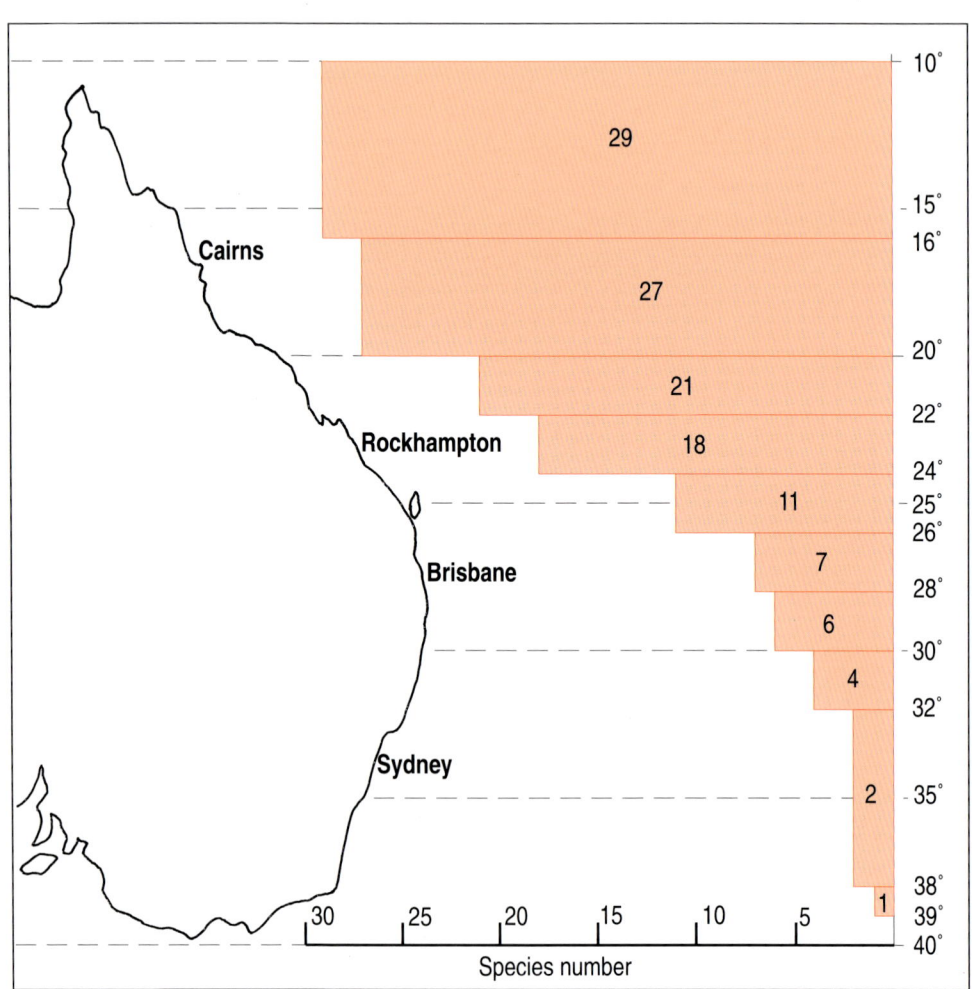

Fig. 3. Diagram of Australian east coast with number of mangrove species according to latitude.

Cape York in the extreme north and Bowen in central Queensland (or from 11° S to 20° S). (There is no universal agreement as to the precise number because of differences of opinion as to what constitutes a true mangrove. The number is about 30 species.)

The tourist industry of tropical Queensland revolves around the splendours of the Great Barrier Reef – and rightly so. However this vast coral barrier system is of vital importance to the coastline behind it: the mighty walls of coral protect the coast from the ocean's swells, and thanks to the mud deposited by numerous rivers, an enormous area of coastal and riverine habitat has been colonised by mangroves. It is estimated that there are some 4,600 sq km of mangrove forest in Queensland. They receive little attention from the tourist trade and yet anyone flying into a coastal town such as Cairns can hardly fail to be intrigued by the seemingly endless mangrove-lined mud flats, rivers, bays and creeks that are visible from the sky.

As one moves south along the eastern Australian coast the number of mangrove species inexorably declines. By the time one has reached Brisbane on the southern Queensland coast there are only seven remaining. The hardy survivors are the river mangrove, the grey mangrove, the large-leafed orange mangrove, the yellow mangrove, the sandy mangrove, the stilt-rooted mangrove and the cannonball mangrove. But this is the southern limit of most of their ranges. South of 30° latitude only the river mangrove and the grey mangrove survive. The southern limit of the river mangrove is at Merimbula in New South Wales (36° 53' S). The grey mangrove's southern limit is Corner Inlet, Victoria (38° 45' S). This is considered to be the northern limit of winter frosts.

When one considers other parts of the eastern region the situation is very similar: the number of species declines sharply the further one moves from the tropics. In New Zealand a variety of the grey mangrove penetrates furthest

south: it occurs near Auckland at a latitude of 37° S. Again, other varieties of the grey mangrove penetrate farthest into the cooler regions than other mangroves where they occur. There are at least half a dozen sub-specific varieties of the grey mangrove listed by various authors, and if they do indeed represent a single species then this must be the most widespread of all mangroves.

The north-western limit of the Indo-Pacific mangrove flora is the North African/Arabian region. It is impoverished in mangrove species. Four species have been reported from there: a variety of the grey mangrove, a prop-rooted mangrove (*Rhizophora mucronata*), the large-leafed orange mangrove and the yellow mangrove. As in the south, it is the grey mangrove that penetrates farthest – reaching the Sinai Peninsula at latitude 28° N. Again, occasional winter frosts are thought to be the barrier that the grey mangrove cannot cross.

In the western region there is a parallel situation. Here it is the very similar black mangrove that demonstrates the greatest temperature tolerance. In the Florida Keys the red mangrove is usually dominant. However, further north it dies out and the black and white mangroves survive. The black mangrove penetrates farthest north: to 29° N in the marshes of Louisiana in the Gulf of Mexico. On the eastern coast of Florida the black mangroves have been shown to halt where the winter temperatures can fall to -4°C. The similarity in overall structure of the black and grey mangroves, and their similarly outstanding eurythermic properties (their ability to tolerate a wide range of temperature) have led scientists to speak of the two as vicariants - originally having belonged to a single species that was divided by some natural event.

Although mangroves are adapted to live in the intertidal region they grow taller in areas that are subject to fresh-water influence, from rainfall, rivers or both. Thus although they survive salt-water immersion, they can be considered to be stressed because they do better in brackish conditions.

Opposite: A coral reef that reaches to the surface can blunt the ocean swells that hit it so that calm waters occur on its inshore side. Mangroves can become established in the lagoon thanks to the sheltering effect of the reef. Here, on a Papua New Guinea reef, a juvenile batfish (Platax pinnatus) hovers in a coral cave.

The grey mangrove is the most temperature- and salinity-tolerant of all mangroves. There are numerous varieties of this species and it is probably the most widespread of mangroves. In the northern Red Sea at Ras Mohammad (left) grey mangroves are the only species capable of surviving the harsh conditions. No fresh water reaches them; they cling to a thin layer of soil barely covering coral rock. Furthermore winter temperatures can drop alarmingly to around freezing point. It should be noted that although they are only low shrubs, the surroundings are plantless desert. The grey mangrove exhibits considerable plasticity of form: in the Northern Territory (right) a grey mangrove that receives a regular supply of fresh water from a nearby stream finds itself in optimum, brackish conditions and grows into a magnificent tree.

In very broad terms, correlations between climate type and mangrove occurrence have yielded the following results:

a) the vast majority of the world's mangal (90 per cent or more) is found in regions that are warm, humid and do not suffer cold winters.

b) Mangroves occasionally occur in sub-humid climates such as Kenya, India and Venezuela.

c) Mangroves are rare in semi-arid conditions, although exceptions include the Indus Delta, Ecuador and the Northern Territory of Australia.

d) Mangroves are very rare in arid climates – exceptions include the Egyptian and Ethiopian coasts – but they can occur in arid areas subjected to winter rains such as the Persian Gulf and the Gulf of California.

The settings in which mangroves appear

Mangroves are found in a variety of sheltered intertidal habitats. They require low energy sites where the silt substrate they need can be consolidated. They are not found on the exposed shores of coastlines subjected to powerful wave action, although they can occur close by: for example the development of a sand bank may allow mangroves to grow on the sheltered landward edge.

Mangroves can be found fringing protected shores, bays, estuaries and deltas and lining the banks of rivers. These are land-based or terrigenous settings. They can also occur on various limestone platforms originally deposited by coral-reef-building animals. These are termed carbonate settings. Terrigenous settings have been classified into the five main types shown in fig. 4.

In the first type the tidal range is low. Deltas are formed by river-deposited sediments, which spread out across the shallow seabed and build new areas of land. The central feature of the delta is the fresh-water river. Salt-tolerant plants are not expected here. However, where there are abandoned deltas across the

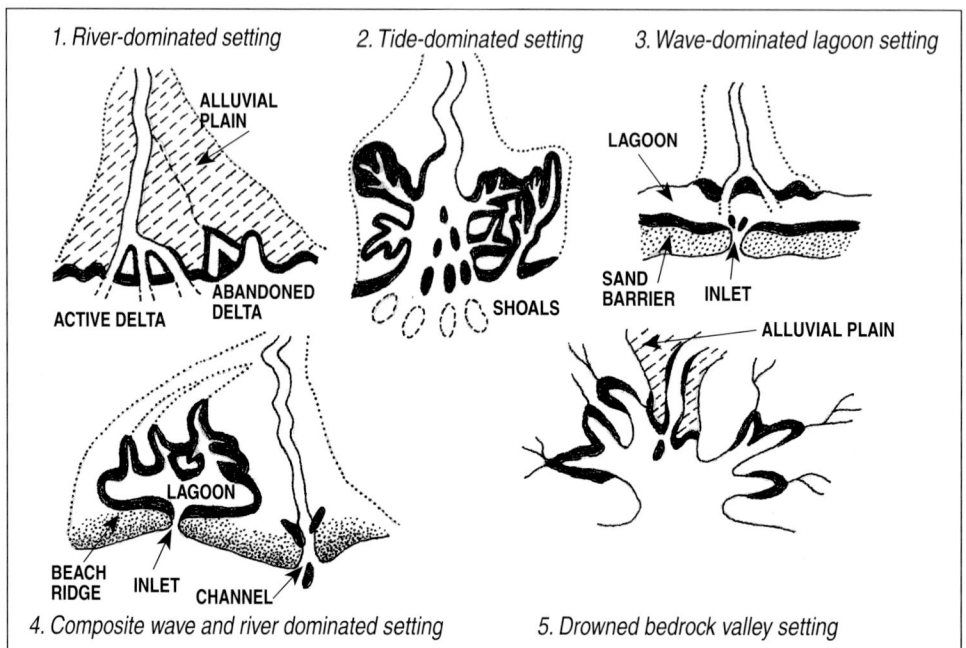

Fig. 4. Settings in which mangroves develop (from Thom, 1984).

alluvial plain mangroves can occur. Wave action softened by the sediment plain deposited by the river can build chenier plains beyond the delta that can accommodate mangrove development. Sediment is brought down by the river and deposited at its mouth to be reworked by currents and tide.

The second setting characterises a riverine coast subjected to strong tidal influence. Wave power is low due to intertidal shoals. In such environments where the sea level has been stable for several thousand years, sediment accretion eventually reaches the height of the highest spring tides. Mangroves can flourish

Table 3. *Approximate areas of mangrove forest in major mangrove areas of the world (in hectares)* (From Wenban-Smith, 1993).

Asia

Bangladesh	410,000	India	356,000	Indonesia	4,251,000
Malaysia	630,000	Myanmar (Burma)	517,000	Pakistan	249,500
Philippines	400,000	Sri Lanka	120,000	Thailand	269,000
Vietnam (S)	370,000				

Oceania

Australia	1,162,000	Papua New Guinea	200,000

West Africa

Angola	110,000	Cameroon	306,000	Gabon	250,000
Guinea	223,000	Guinea-Bissau	236,000	Nigeria	3, 238,000
Senegal	169,000	Sierra Leone	250,000		

East Africa

Madagascar	326,000	Tanzania	134,000

Eastern Americas

Bahamas	233,000	Cuba	626,000	Honduras	117,000
Mexico	660,000	USA	280,594	Brazil	250,000
Surinam	115,000	Venezuela	674,000		

Western Americas

Mexico	660,000	Panama	298,000	Colombia	501,000
Ecuador	182,000				

both along the numerous creeks and also on the offshore sediment islands.

The third setting shows how mangroves can develop in areas with high wave action. The wave action has deposited sediment from elsewhere which acts as a protective barrier to the river mouth. A lagoon results. The river discharge is low and the lagoon does not become filled. Mangroves can develop in the numerous kinds of feature that can occur in such a lagoon.

Setting number 4 shows the conditions in which mangroves can develop despite high wave energy and heavy river discharge. The wave action redistributes coarse sediment deposited by the river into ridges that parallel the shore. The lagoons thus formed have shallow alluvial plains as their landward edge. These areas are readily colonised by mangroves.

The last terrigenous setting is a river valley that has been drowned by rising sea levels. Sheltered river mouths here may deposit sediment that is not swept away by wave action, or wave action may have deposited fresh sediment to form a tidal delta at the mouth of the river mouths.

Near Wyndham in Western Australia, mangroves fringe the numerous creeks of an extensive floodplain as well as lining the sheltered coastline.

In carbonate settings, mangroves can either establish themselves directly on the structures formed by the calcium carbonate depositions of marine animals or on substrates of such reworked material deposited by tide or wave action. Three such settings have been classified.

In setting 6, mangroves can become established on the carbonate platforms of low-energy coastlines where marl or peat has been desposited to provide a substrate for their roots to grow in.

In setting 7, an offshore sand barrier that hinders wave action can allow mangroves to grow behind it

Mangroves cannot develop on exposed coastlines subjected to even moderate wave action. However the build-up of sand barriers between the sea and creek/lagoon areas can provide the shelter mangroves need, as here in Darwin Harbour in the Northern Territory.

on a carbonate platform. The material that the roots grow in may be sand or a mixture of sand and peat, although peat accumulation derived from the mangroves themselves is likely to be modest.

Finally, in setting 8, sheltered or low energy bays, mangroves can become established on the drowned remains of prehistoric reefs.

Mangal can become established in a wide range of geomorphic settings. Scientists working on the four most common New World mangroves (the red, black, white and buttonwood mangroves) have classified mangrove communities into the six basic types shown in figure 3.

Flooded river valleys in Darwin Harbour provide a sheltered setting for mangroves because the sediment deposited is not washed away by strong outward-bound currents; indeed extra loads of sediment can be deposited by wave and current action, providing a wide variety of settings for mangroves.

Mangrove soils

Mangroves need soft sediments for their root systems. A variety of substrates can be colonised: sands, muds, clays and silts. Although they can be established on rock, there must be pools of soft sediment that their roots can enter. Anyone who has explored a wide range of mangrove forests will have noticed the variety of soil types in which they grow. In one place the ground underfoot will be relatively firm, perhaps a coarse sand. In another it will be a clinging clay.

In yet another it will be a soft silt that is impossible to walk on. Dig a little under the surface and the mud turns black and exudes the stench of rotten eggs.

One of the remarkable things about mangroves is their ability to grow in thoroughly unpleasant types of soil. The most conspicuously unusual feature of many mangroves is their strange root adaptations and this is in response to the type of soils in which they are found. I outline below some of the problems facing plants growing in intertidal soils, but if you are not interested in the chemical and biochemical processes involved, you need only know that the oxygen-lacking salt-water-saturated mud in which mangroves grow cannot provide the roots with the oxygen they need for their development. Aerial roots are a response to the problem: atmospheric oxygen is taken into the roots through the aerial structures when the tide is out.

Normal well-aerated terrestrial soil will contain air-derived oxygen. It will also contain a population of bacteria that respires aerobically. Waterlogged intertidal soil lacks enough oxygen for a significant number of aerobic bacteria to survive. Anaerobic

In sheltered bays such as here in the Bahamas, mangroves can gain a foothold on a substrate of coral rock if there is sufficient sand deposition to allow their roots to become established.

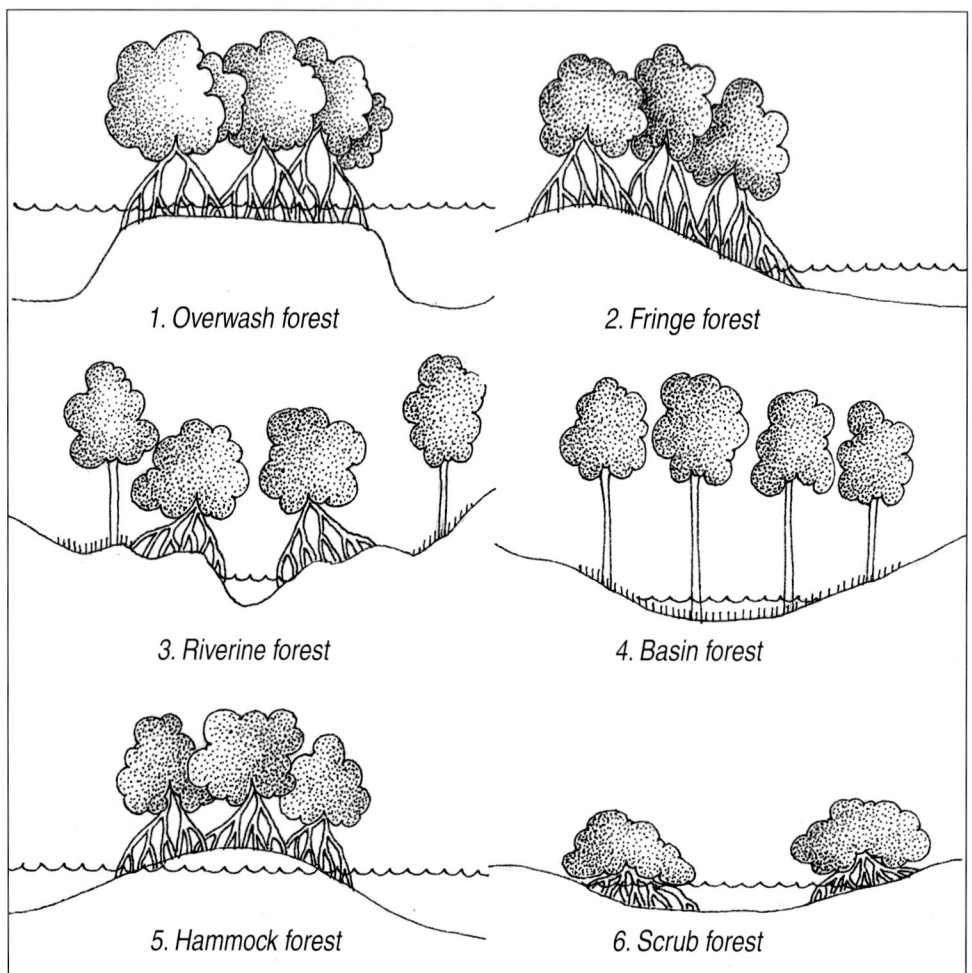

1. Overwash forest

2. Fringe forest

3. Riverine forest

4. Basin forest

5. Hammock forest

6. Scrub forest

Fig. 5. The types of mangrove community (from Day et al 1989).

bacteria – those that do not need oxygen to respire – will flourish provided the soil contains other oxidant sources for respiration. Anaerobic bacteria can be divided into two types: the obligate and the facultative anaerobes. The obligate anaerobes cannot survive in the presence of oxygen. Facultative anaerobes (the dominant type in mangrove mud) have the capacity to respire aerobically when oxygen is present, and switch to anaerobic respiration when it is lacking. If there is limited oxygen in the soil then it will rapidly be used up in aerobic respiration. From then on the facultative anaerobes will switch to anaerobic respiration.

Dig a little below the surface of mangrove mud and the soil turns black and emits a rotten smell. Mangrove soils lack oxygen and are rich in sulphur compounds.

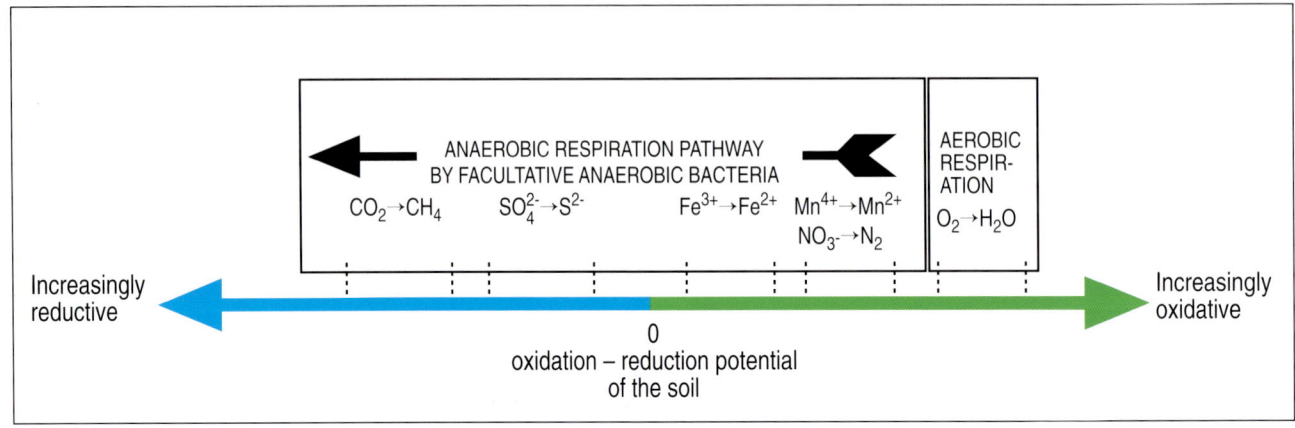

Fig. 6. Chemical processes by facultative anaerobic bacteria of increasingly reductive soils. From Boto 1984 (simplified).

There are several different chemical steps that will make the soil ever more anaerobic. Progressively more reductive processes occur (if the appropriate raw materials are present) that reduce the nitrate ion, manganese (IV) and iron (III). Thereafter sulphate derived from sea water will be reduced to sulphide and carbon dioxide to methane (marsh gas). It is these last two steps in severely anaerobic soils that produce the characteristic stench when mangrove mud is disturbed.

There are subtle problems facing plants that grow in anaerobic soil. They need nitrogen for growth; a lack of sufficient nitrogen will stunt their development. However nitrogen can only be appropriated by plants when it is incorporated in inorganic structures such as ammonium and nitrate ions. Anaerobic soils often have a thin film of aerobic soil on their upper surface. Ammonium from the anaerobic zone will diffuse into the aerobic zone where it will be oxidised by microbes into the nitrate ion. This will then diffuse back into the anaerobic zone where it will be reduced to nitrogen gas and nitrous oxygen which will

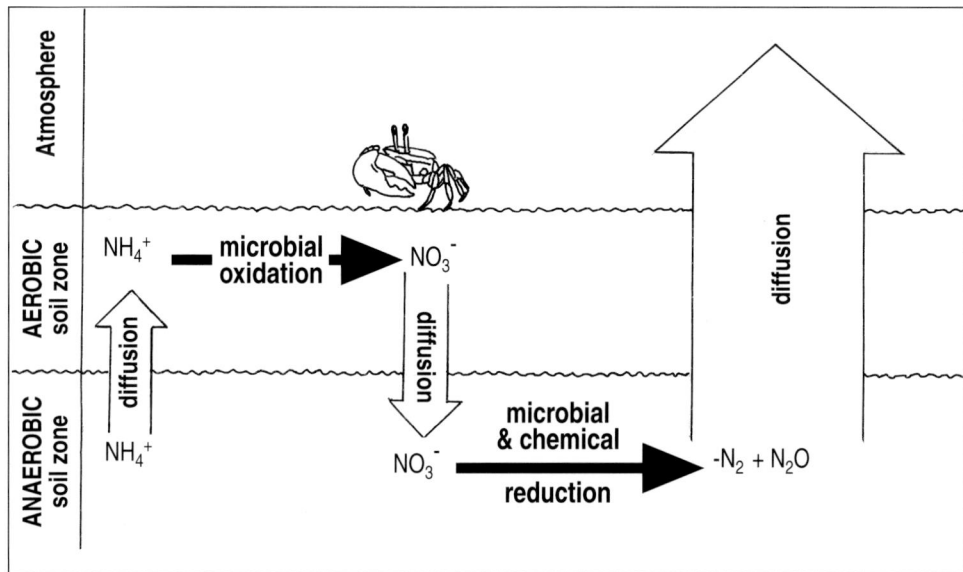

Fig.7. *Diagram showing denitrification process of anaerobic soils: how mangrove soils lose nitrogen. From Boto 1984.*

diffuse out of the soil into the atmosphere, thus depriving plants of a crucial raw material for their growth. In other words nitrogen, a crucial component for plant development tends to be lost spontaneously from anaerobic soils.

There are other challenges facing a plant growing in anaerobic soil saturated with salt water. The organic matter of the dead plants can decompose very slowly in the anaerobic mud. Recycling becomes slower and slower. Over a long period, an imbalance can result between the amount of nitrogen and phosphorus needed by growing plants and the amount available in the soil: the slow rate of decay means that progressively more is trapped in rotting material and less is available to the new generation of developing plants.

Another problem is that certain metals that are toxic to plants become progressively more soluble in an acidic environment and mangrove mud often tends towards the acidic. Furthermore there is the problem of the extreme toxicity of the sulphide ion in the soil. In highly anaerobic soil sulphide will combine with appropriate metals and trap them in an insoluble form that cannot be utilised by plants. There are also certain kinds of anaerobic saline soil that produce sulphuric acid when re-aerated. Tampering with mangrove areas (for instance draining them for agricultural purposes) can destroy the plant-carrying capacity of the soil.

The physical characteristics of mangrove soil will vary greatly from one location to the next. For example the size of the sediment particles – from coarse sand to silt – will have profound effects. Particle size will affect the drainage capacity of the soil. It will influence the amount of evaporation of water when the soil is exposed to the atmosphere,

The ability of mangroves to grow in the oxygen-lacking, salt-water-saturated, sulphur-rich muds of the intertidal realm is nothing short of remarkable. They are the primary producers of the estuarine waters in which they occur: they harness the energy of the sun to build organic compounds out of carbon dioxide and water. Here a young stilt-rooted mangrove plant struggles into life in the Northern Territory.

thus affecting salt build up. Sediment size will also be an important factor affecting the degree and rate of diffusion of atmospheric gases into the soil and determine the depth of the aerobic upper surface, which tends in any case to be low in oxygen and high in carbon dioxide and hydrogen sulphide - little wonder then that mangroves have developed aerial root systems to aerate their roots.

In addition to the effects the soil has on the mangal, there is also a very important point to be made about the effect of the mangal on the soil. Once established a mangrove forest is an almost impenetrable barrier of tree trunks, prop roots, pneumatophores, seedlings and saplings. This dense forest of obstacles becomes a sediment trap. Both the sediment carried to the mangal from elsewhere (allochthonous sediment) and the material derived from the mangrove ecosystem itself (autochthonous sediment) will be slowed by this barrier and tend to settle out on the bottom. Furthermore the extensive root systems that are just below the surface of the substrate hold the sediment in place.

Do mangroves build land?
There are two longstanding myths concerning mangroves and their sediment-trapping ability that need to be examined. The first is that mangroves extend shorelines. Mangroves on the seaward edge of the mangal (in the New World this is typically the red mangrove) trap sediment. This itself becomes available for colonisation by new mangrove seedlings. When these grow they trap more sediment and so on. Thus in areas of shallow water mangroves have been thought to extend the coastline. However, there are areas where mangroves have failed to colonise such shallow areas over extensive periods. Furthermore if mangroves genuinely could extend shorelines then planting them in areas of coastal erosion would prevent such erosion. This does not appear to work. The error is that earlier investigators studied mangrove forests in areas of naturally

Hundreds of snowy egrets (Egretta thula) *feed on extensive mudflats in Trinidad's Caroni Swamp. The red mangrove forest acts as a sediment trap so that mud is not readily washed out to sea.*

occurring coastline accretion and assumed that the accretion of sediments was due to the mangroves themselves. Mangroves will naturally colonise suitable habitats formed by such accretion, but their establishment should be seen as a response to the ever-changing intertidal landscape rather than a cause of it. Once established they will trap vast amounts of sediment, but this does not mean that they are extending the shoreline itself.

Where heavy amounts of allochthonous sediment are deposited to form new intertidal mudflats, the response by colonising mangrove seedlings can be rapid.

Below the level of low tide extensive areas of sea grass can abut mangroves. This domain is heavily dependent on organic nutrients derived from the mangroves. Dugongs (Dugong dugon) feed on various types of sea grass and are fairly common in the sea grass areas of northern Australia which are contiguous with mangroves. A mother and calf are pictured here. In the mangrove creeks of the tropical Atlantic several species of similar-looking manatees (Trichechus spp.) are found. All these varieties of sea cow are endangered over much of their range due to destruction of habitat, hunting, drowning in fishing nets, being hit by boats and pollution.

For example in Sumatra the seaward advance of mangroves has been measured at 125 m per year and in Java at as much as 200 m per year. (While mangroves will rapidly colonise areas made available by allochthonous sediments they cannot establish themselves in areas receiving constant layers of fresh sediment as their root systems are too slow to respond to the ever-changing levels.

The other related myth is that mangroves raise the height of the land. Leaf litter, parts of the trees and organic debris deposited by the tide will tend to be trapped in the mangal and form peat, and it was supposed that this would

In the shallow waters of Fergusson Island, Papua New Guinea, stony corals can be seen in the foreground growing right up to a thin belt of large-leafed orange mangroves (with the buttressed trunk bases) and some prop-rooted mangroves. Stony corals can develop here because of the minimal amount of sediment reaching them from the mangroves. The large-leafed orange mangrove is often found on the seaward edge of small, isolated Indo-Pacific mangrove communities that lack the typical seaward edge flora such as the mangrove apple and the grey mangrove. Also visible here are some coconut palms growing on elevations of land above the reach of the tide.

eventually rise above the highest tides, enabling other non-mangal plants to become established. Indeed this model has long been thought to apply to the red mangrove: its very success seals its own fate as the peat build-up raises the land above the height required by the species.

However, although peat is relatively long-lasting when it is saturated by water, when it is exposed to atmospheric oxygen it quickly decomposes. In other words any peat that did rise above the tidal level would not persist for any length of time and could not possibly form new land.

Mangroves and neighbouring communities

Mangroves grow best in low-energy intertidal areas. They are often bordered by other important shallow-water communities, the most common being sea grass beds. Patch reefs of coral (coral reefs developing on anomalies of the sea floor such as areas of exposed rock) and even fringing reefs (coral reefs that parallel the coastline and are close to the shore) can also be found next to mangroves just as mangroves can develop on sand islands formed behind coral reefs. But this does not happen to any great extent. Coral reefs require rather different conditions from mangrove swamps. They are beneath the intertidal range, although a few corals can survive a short amount of atmospheric exposure at extreme low tide by secreting a mucous coating that slows desiccation and may even act as a sun screen. Coral reefs tend to be found in high-energy areas, and it is the ability of the coral wall to absorb the sea's energy which allows mangroves to develop behind it. Corals, however, need fairly clear water to allow the photosynthesising algae in their tissues to function, in addition to which heavy sediment loads will smother and kill the corals. The more extensive a mangal the more turbid the waters leaving it are likely to be. So local coral development will be stunted, and vice versa: areas with spectacular coastal reefs usually lack extensive mangal areas.

Having discounted the land-building myth about mangroves, it might be worth mentioning another, rather more plausible theory: that corals and sea grasses can build up the land, leading to a succession from corals to sea grass beds and on to mangal. This theory is not based on the fact that coral rock can provide a foundation over which mangroves can grow, but rather on the fact that certain kinds of coral reef, as they develop, form a lagoon between the reef and the shore. If the land-mass on which a coral reef develops slowly submerges, and the reef can keep pace by growing upwards towards the surface, then the

On Fergusson Island in Papua New Guinea, mangroves fringing the coastline give way to tropical rainforest as the land climbs above the height of the tide. This proximity between habitats means that many species of rainforest animal (though few plants) can enter the mangroves.

area between the reef and the retreating shoreline will become flooded: a lagoon will form. The lagoon will become a sediment trap for the weathered and (in the case of coral-browsing animals) excreted sand derived from the broken-down coral. This shallow area can be colonised by sea grasses, which will increase the preciptiation of coral sand by slowing the water movements. The root systems of the sea grasses will bind the sand so that more and more accumulates and the area becomes progressively more shallow. Eventually parts of the lagoon will be shallow enough to be colonised by mangroves.

Leaf litter, peat and detritus

Despite the harshness of the environment in which mangroves are found, they have been shown to be one of the most productive of ecosystems in terms of both their gross primary production and their leaf-litter production.

Fundamental to an understanding of life on Earth is the idea of primary production. Plants take simple inorganic constituents and combine them into complex organic substances. The raw material for the synthesis is atmospheric carbon dioxide; the nutrients are inorganic ions such as phosphate and nitrate; the energy for the process is provided by the sun. The process is termed photosynthesis, and the products are complex organic substances – carbohydrates, fats and proteins. The gross primary production is all the organic matter assimilated by an autotroph (an organism that uses atmospheric carbon dioxide as its exclusive or main carbon source for synthesis). However some of the carbohydrate is used by the plant to fuel its own life processes. The remainder is mostly used for the production of new plant tissue. This is termed the net primary production.

Plant tissue is the food source of herbivores that are themselves preyed upon by carnivores. The mangrove plants are the dominant primary producers of the

Above: At high tide a school of zooplanktivorous fishes
(Jenkinsia lamprotaenia) *find shelter from predators*
amonst the pneumatophores of a black mangrove tree in the
Bahamas.

Right: The osprey (Pandion haliaetus) *is a common fish*
predator of mangrove swamps. Here in the Northern Territory
an osprey grips its prize.

intertidal tropical and subtropical areas in which they occur. (In temperate climes the mangrove ecosystem is replaced by the salt marsh community. This grass community occupies the equivalent niche: sheltered intertidal areas.) There will often be other primary producers in the vicinity of the mangal – for example beds of sea grass or bottom-dwelling algae. And above the tidal range terrestrial plants are likely to border it. The dependence of animals on the mangal will become apparent later; suffice it to say at this stage that the significance of the mangrove ecosystem can hardly be overestimated.

The primary production in the open sea is performed by the plant plankton (phytoplankton). The phytoplankton is consumed alive by animals and therefore termed a grazing food web. On the other hand much of the plant material of inshore waters – including the vast bulk of mangal matter – dies and then decomposes. This decomposing material is termed organic detritus and it is the basis of the detrital food web. Mangal detritus is derived from every decomposing part of the dead plants, including the leaves, branches, roots and failed seedlings. Much of this material will be eaten by animals within the close vicinity of the mangal, but the great nutrient soup of detritus that is constantly seeping out of the mangal into the inshore waters feeds numerous other creatures as well.

Where the mangal borders a calm estuary, or the flow to the sea is slowed by natural barriers, a rich and complex inshore food web of plants and animals can become established that is largely dependent on

The tiger shark (Galeocerdo cuvier) is an apex predator of inshore waters throughout the tropics and sub-tropics. It will enter very shallow water in order to feed and will eat an enormous range of food items. (Note the red mangrove hypocotyl on the bottom in front of the shark.)

mangal detritus for its sustenance. Beneath the bottom there will be an anaerobic food web of bacteria, protozoans and worms breaking down the sunken detritus. In the water column plant plankton will be consumed by animal plankton (zooplankton) which is itself eaten by schools of zooplanktivorous fishes. On the bottom will be filter-feeding mussels and oysters. Crabs, crayfishes, starfishes, shrimps, prawns and snails will scavenge across the seabed. Bottom-dwelling bony fishes will graze on the plants and animals of the sea floor. Stingrays will forage for crustaceans and molluscs in the muddy substrate.

At the apex of the food web will be a variety of predators adapted to eat a wide range of food items. These will include birds such as sea gulls and sea eagles. Large generalised predators amongst the bony fishes will include barracudas and (in the New World) tarpon. Indeed the tarpon (*Megalops atlantica*) has a special adaptation to allow it to live in oxygen-deficient, even stagnant, inshore

Table 4. *A comparison of the net production of dry organic matter in various ecosystems in grams per square metre per year -*

Open ocean (average)	100 g/sq m/yr
Rich coastal waters (average)	600 g/sq m/yr
Rainforest (West Indies)	6000 g/sq m/yr
Tropical mangrove	2500-3600 g/sq m/yr
Coral reef (Pacific Ocean)	4900 g/sq m/yr

From Lear and Turner (1977).

waters: there is a connection between the swim bladder and the gullet so that the fish can breathe by gulping air. Amongst the cartilaginous apex predators of the inshore/estuarine ecosystem there are sharks which are especially adapted to the murky shallows, like the bull shark (*Carcharhinus leucas*) which will eat almost any animal it can catch. It can also penetrate rivers far upstream.

Scientists constantly emphasise the vital nutritive rôle of mangroves for neighbouring populations of both animals and plants. Nevertheless it is extremely difficult to quantify their significance. As a negative correlation removal of mangrove forests can lead to the collapse of offshore fishing industries. For example the commercial prawn fisheries off Ecuador have suffered a disastrous collapse that coincides with the clearing of considerable areas of mangrove,

A sample of animal plankton from inshore waters shows a variety of small animals including the larva of a decapod crustacean (with the large eyes), calanoid copepods and the long, thin bodies of arrow worms.

ironically for the building of artificial prawn ponds designed to farm the very species of prawn that were previously abundant offshore. And yet the prawn farmers cannot now adequately stock their ponds with these prawns because pond construction has devastated the offshore population. Sadly this ill-informed destruction of the mangrove habitat for a failed enterprise is being repeated time and time again around the world. Where once there was mangal now there are failed prawn ponds.

A more positive correlation between mangal and offshore fishery yield can be found in Australia's Gulf of Carpentaria. Over a period of 15 years scientists compared the prawn catch of commercial fishermen with periods of rainfall. They found that the best catches were made during periods of heavy rain.

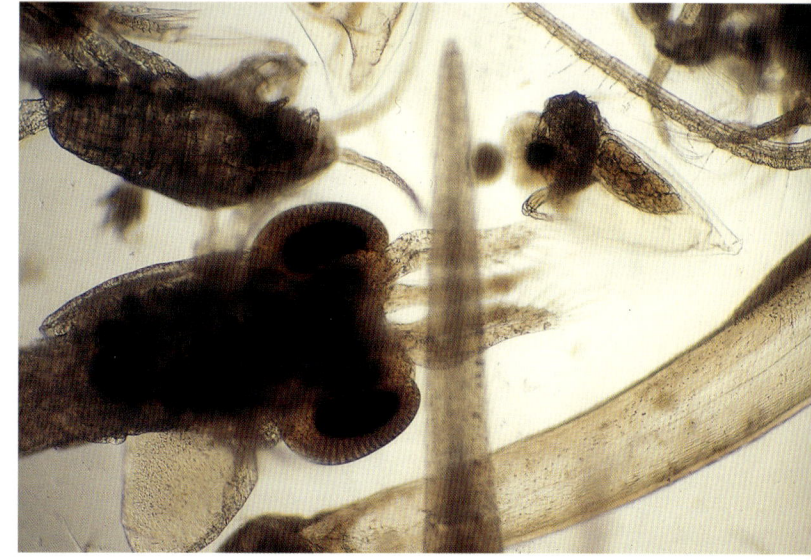

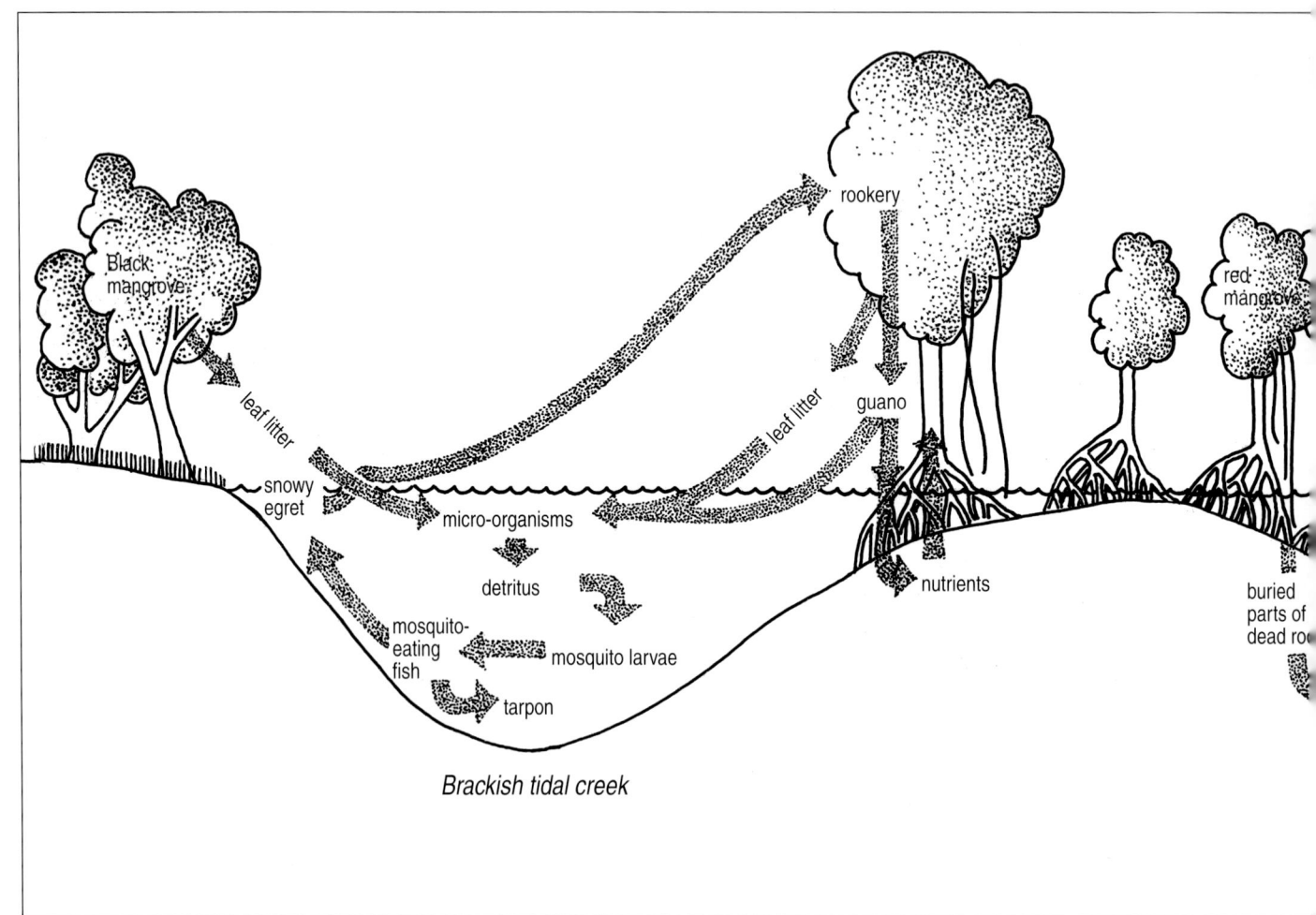

Brackish tidal creek

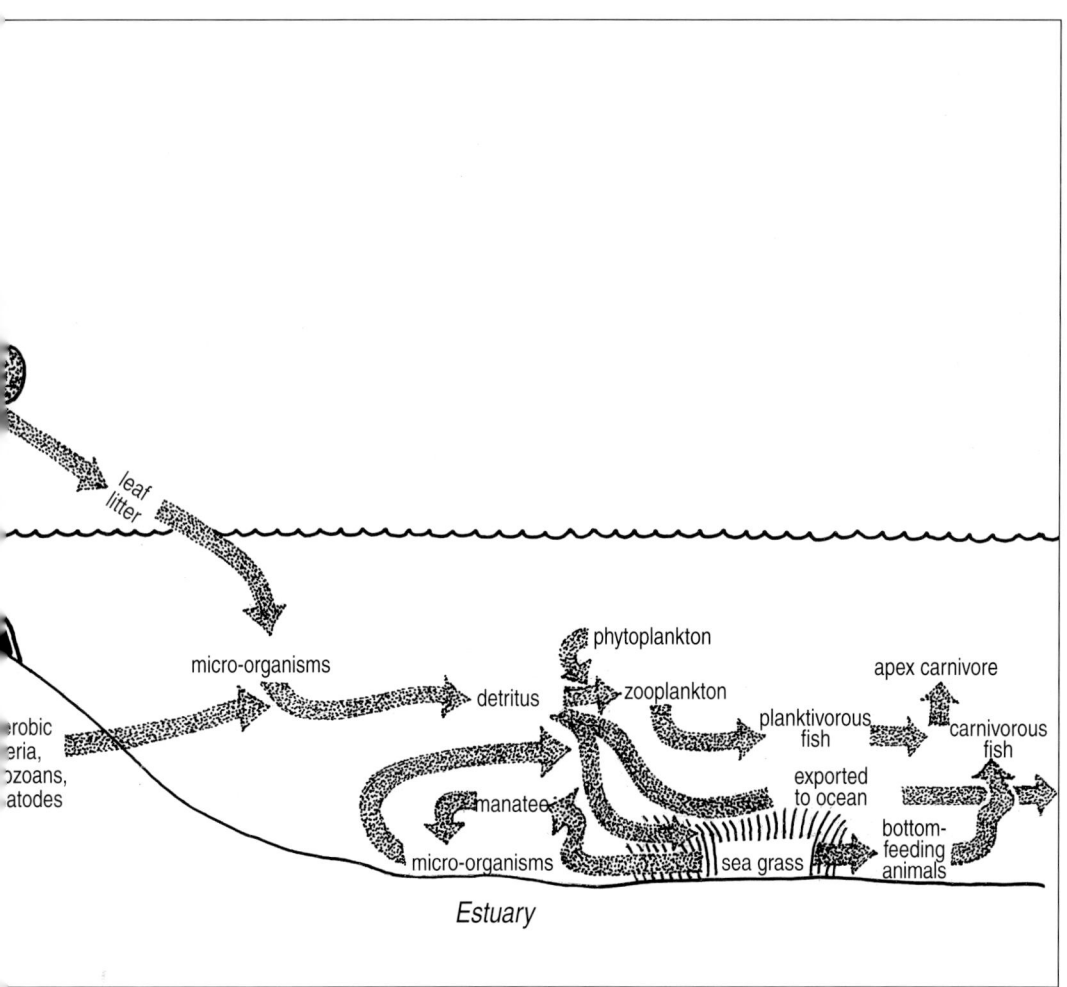

leaf litter

micro-organisms

aerobic
bacteria,
protozoans,
nematodes

detritus

phytoplankton

zooplankton

apex carnivore

planktivorous
fish

carnivorous
fish

exported
to ocean

manatee

bottom-
feeding
animals

micro-organisms

sea grass

Estuary

Fig. 8. Simplified diagram of the detrital food chain in a mangrove-fringed tidal creek and estuary in the Caribbean.
Adapted from Boaden and Seed (1985) and Day et al (1989).

They inferred therefore that increased rainfall benefited the local mangroves by reducing the salinity. This led to an increase in primary production by them, and one of the results was the increased production of leaf litter which enriched the detrital food chain. Many species of commercially valuable prawns (Penaeidea) use inshore mangal-fringed waters as nursery grounds for their post-larval development.

This raises the question of how significant a contribution the organic matter (as detritus) derived from local mangroves makes compared to the total input of organic material derived from outside the estuarine environment. This will depend on various factors such as the amount of tidal energy available for exporting detritus and the amount of leaf litter produced by the mangal in question. Work done on mangrove-fringed estuaries in southern Florida showed that in one (Rookery Bay) 83 per cent of the organic matter carried into the estuary was derived from the mangroves while in the other (Fahkahatchee Bay) it was 52 per cent.

The organic input into the estuary from mangroves supports a high population of secondary consumers (zooplankton) that also feeds on the major primary producers of the estuary – the phytoplankton. It might be argued that increased detritus input from the mangroves

Termites are an important component of mangroves because they feed on dead wood and therefore accelerate its breakdown and hasten the rate at which organic nutrients can be recycled within the ecosystem. This termite colony was photographed in Kilgwyn Swamp in Tobago. The termite in the centre with the pale head is a worker. The termites with the dark, nozzle-shaped heads squirt noxious chemicals at intruders (usually ants) to defend the colony. Their presence identifies this colony as belonging to the Nasutitermes *genus.*

could cause a boom in the zooplankton population which also feeds on the phytoplankton. Certainly excessive detrital input can make the water so murky that light cannot penetrate and photosynthesis by the estuary's primary producers virtually halts. However it may be that the primary producers of the estuary also benefit from the organic matter derived from mangroves: research has suggested that increased primary production by estuarine phytoplankton coincides with increased input of detritus from mangroves.

Researchers working in the Florida mangroves have attempted to measure the amount of leaf litter actually entering the estuarine habitat from the mangal. Their experiments demonstrated that some 800 g of leaf dry matter falls on each square metre of ground per year. Ninety-five per cent of this dry matter is not consumed in its original form (dead leaves) and only about two per cent is stored as peat. The dead leaves will become saturated with water and subjected to structural breakdown due to chemical, mechanical (friction, wave action) and biological attack. Of this remaining organic material (over 90 per cent of the original) about half is eventually consumed by bacteria, fungi and detrital feeders within the mangal and the rest is washed into the estuarine environment.

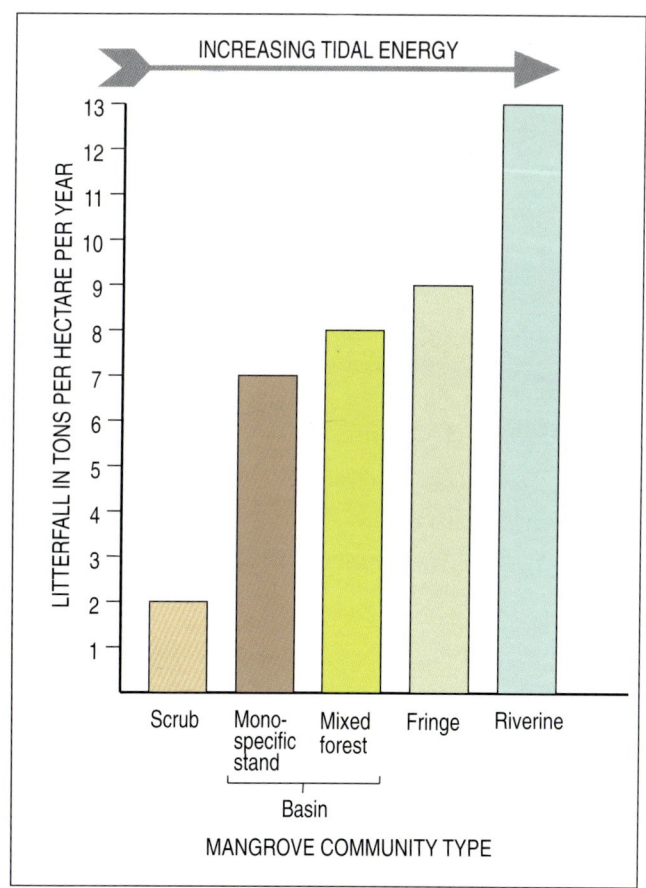

Fig. 9. *The relationship between litterfall, tidal energy and mangrove community type. From Twilley (1988).*

Is the measurement of the productivity of this Florida mangrove typical of others? Far from it. Values for leaf litter production in mangroves vary enormously from 1.2 metric tons per hectare per year in Florida mangrove scrub to 23.4 metric tons per hectare in a Malaysian mangrove. One of the reasons for this wide range is the height and frequency of the tide. For example frequent tidal immersion will reduce the salt build up owing to evaporation when the tide is out. It may also help flush toxins such as hydrogen sulphide out of the upper levels of the soil. In fact litter production

An extensive bloom of green algae that has attached itself to the pneumatophores of mangroves in Ludmilla Creek, Darwin Harbour. This unusual sight may be due to pollution – nutrient enrichment from a nearby sewage outfall.

in various mangrove habitats has been shown to correspond to the amount of tidal energy to which the community is subjected. The greater the tidal energy the greater the productivity: riverine mangal is the most productive, followed by fringe and basin mangals, and the least productive is scrub mangal.

The greater the tidal energy at a given site the more detritus will be carried away into the estuarine habitat. Measurements have shown that in a basin forest subjected to low tidal amplitude only about 20 per cent of leaf litter was exported, compared with some 45 per cent in a fringe forest.

While we noted earlier that mangroves are generally rare in arid climates it may be that some of the problems faced in these areas can be offset by high tidal energy. Certainly the considerable tidal range along the Northern Territory coastline of Australia may allow mangroves to overcome the problems caused by the aridity of the region.

It should be noted that while the mangrove plants are the dominant primary producers of this ecosystem, the numerous types of algae that attach to the roots of the trees, to the trunks and also to the mud itself can make a significant contribution to the gross primary production: measurements in New World mangals has demonstrated that this can be as high as 16 per cent of the total.

The process whereby a discarded mangrove leaf is broken down into detritus is a complex one. The variety of rôles played by microbes (organisms less than 0.1 mm in size) has only recently been appreciated. The mangal detritus is infested with bacteria and protozoans that break it down into components the microbes can use. This stage is crucial to many larger animals that lack the enzymes necessary to break down the complex structural components of plants: namely cellulose (the structural component of plant cell walls) and lignin (the other basic structural element, with cellulose, of wood). The microbial biomass eaten by larger animals contains much organic matter that was originally unavailable to them because it was locked up in these indigestible structures. Furthermore the corpses and faeces of the larger animals are broken down by microbial activity and this is incorporated in the microbial biomass, and the nutrients are kept within the estuarine habitat. On the other hand the microbes can actually compete with the primary producers for the available inorganic nutrients.

There are two forms of detrital decomposition: aerobic and anaerobic. The latter, which occurs below ground, is little understood in terms of its significance for total detrital matter but includes the root systems of the mangal as its basic component. The organisms that perform the task of anaerobic decomposition are mostly microscopic but there are also types of worm (nematodes) that are facultative anaerobes. Initial studies in salt marsh communities in Massachusetts suggest that this is a surprisingly productive form of metabolism: carbon production from this area has been reported at 850 g per square metre per year.

Aerobic detrital decomposition has been better studied. The process by which a discarded mangrove leaf decomposes has several steps. Mechanical action by water and friction or collision with other objects will weaken the overall structure of the leaf. It will be colonised by bacteria and fungi. It will gradually break down into smaller pieces which are themselves attacked by shrimps and crustaceans. The decomposing leaf matter is colonised by microbes which convert some of it for their own uses. It is ingested by larger animals, the microbial component being digested, and the remainder excreted as faeces. The faecal material will then be reinvaded by microbes which feed on the remainder of the carbon component. They will be eaten by larger animals and so it will go on, the carbon component of the original leaf matter being progressively reduced.

Scientists have studied different mangroves and their leaves in an attempt to understand better the processes whereby the plant tissue is degraded into detritus. It has been shown that in a mangal in Africa it takes about six months for the plant material to be reduced to organic detritus. Finely meshed bags were used to hold the leaves *in situ* during the experiment and later examination of the bags revealed that they had been invaded by large numbers of peracaridan crustaceans (amphipods, isopods and tanaids) as well as by polychaete worms. Because the holes in the net were smaller than the adult animals in the bag it was realised that the polychaetes must have entered either as eggs or shortly after hatching. The worms were presumably a significant contributor to leaf breakdown, although whether the crustaceans actively contributed or were passive filter-feeders was not established.

The rate of decay of the leaves of the species studied – the grey mangrove and the large-leafed orange mangrove – were quite different. The grey mangrove leaves decomposed faster and released their nutrients more rapidly. Furthermore the large-leafed orange mangrove's leaves were more buoyant: the majority floated for at least a week at varying salinities while those of the grey mangrove mostly sank within a day. After 13 days all the large-leafed orange mangrove leaves in fresh-water had sunk; after 15 days 10 and 15 per cent of the leaves in brackish and sea water-strength solutions of dissolved salt were still afloat respectively. The buoyancy can be understood in terms of the structure of the large-leafed orange mangrove's leaves: they are coated in a thick, impervious cuticle and have large internal air spaces. The large-leafed orange mangrove's leaves are more likely to be lost from the mangal environment than those of the grey mangrove.

A pufferfish (Arothron hispidus) *rests amongst the scattered flowers of a large-leafed orange mangrove in Papua New Guinea.*

An immature little blue heron (Egretta caerulea) *hunts in the shallow waters of Tobago's Kilgwyn Swamp.*

The fossil record

There are considerable problems involved in trying to understand the evolution and distribution of mangroves over geological time. The first problem is to know what fossilised items can clearly be identified as the remains of mangroves as opposed to other plants. For example fossil remains of pneumatophores do not necessarily signify prehistoric mangroves: they could have originated from trees in fresh-water swamp habitats. A more productive method is the analysis of prehistoric pollen and spores. This, the discipline of palynology, is the major method of investigating prehistoric mangrove communities. Although it is often impossible to identify the parent plant at the species level (due to the similarities of intrageneric pollen) palynology can be highly accurate to the level of the genus.

Many of the principles that hold good for a modern mangal can be used to infer the challenges facing the establishment of prehistoric ones. Consider that remarkable adaptation of the modern angiosperm mangroves, the development of vivipary. The problems facing development in salt water are largely circumvented by these higher plants, by adaptations to the seed/seedling stage of development. It seems reasonable to assume therefore that more ancient plant groups such as ferns, which lack the seed habit, did not establish prehistoric mangals. The mechanism of reproduction of the ferns alternates between two very different generations: the sporophyte (what we think of as the fern) and the tiny, fragile gametophyte, which contains half the set of chromosomes of the parent sporophyte. It is highly unlikely that the gametophyte could develop in salt water. While it is true that there are several fern species found in modern mangal (of the genus *Acrostichum*) and although they possess a heightened tolerance to salt water, they are always found on the landward and never the seaward edge. It has been suggested that a prehistoric fern (*Weischella*) formed extensive

The white-breasted sea eagle (Haliaeetus leucogaster) *is a generalised apex predator found in Australian mangroves. It feeds chiefly on fishes.*

mangal throughout the tropics in the early Cretaceous Period (about 130 million years ago), but it is likely that it was confined, like its present-day counterpart *Acrostichum* to the landward edge of the mangal or to areas frequently subjected to fresh water immersion.

There appears to have been no shortage of suitable habitats in the geological past for mangrove colonisation. Indeed much of the equatorial region was composed of massive bodies of land gradually sloping into shallow seas. However, we must be cautious when attempting to ascertain where prehistoric mangrove communities were from fossilised parts of mangroves such as pollen – ocean currents will carry and deposit propagules and microspores far beyond their viable range.

Two questions that can be asked of the fossil record are: how far back in geological time do present-day mangroves stretch, and what is the evidence for extinct mangroves that predate the appearance of modern mangroves?

In answer to the first question, the mangrove palm seems to be the most ancient of extant mangroves, with its origins at the end of the Cretaceous or the beginning of the Tertiary (about 65 million years ago). The pollen spores are so characteristic that they can be identified at the species level. They look like two spike-covered hemispheres joined together by their flat surfaces. Because the pollen adheres to animals (fruit flies are thought to be the main pollinator) it is thought that the distribution of fossil pollen accords well with the distribution of prehistoric *Nypa* mangal; fossilised fruit has also been collected. The evidence suggests that the mangrove palm whose range is now confined to the region from Sri Lanka through Malaysia and Indonesia to New Guinea and Australia and north as far as Japan, previously extended to Brazil, Europe and Africa.

Amongst other extant mangroves, the prop-rooted mangroves and *Pelliciera* have been recorded as far back as 30 million years ago. Pollen from the peg-rooted mangroves has been found in early Miocene deposits (24 million years ago) and pollen from pencil-rooted mangroves is known from the mid-Miocene (about 15 million years ago).

The oldest suggested angiosperm mangal is from the late Cretaceous (the Dakota Sandstone of Kansas) with a date of about 75 million years ago. Indeed it has been suggested that modern flowering plants evolved from coastal proto-angiosperms and that the coastal habitat was crucial to their dispersal.

To answer the second question we must go back 295 million years to the

Upper Carboniferous Period. This was a time when huge swamp forests covered the land. It was the age of the lycopods: huge trees 30 m or more in height, whose fossilisation formed the majority of the world's coal beds. Today's club mosses are related to this extinct group. What is significant is that club mosses need *fresh* water in order to reproduce, so presumably the same was true of the lycopods. Researchers trying to find evidence of ancient mangrove swamps have looked at several coal faces from this period in Iowa where very different plant communities are found. These swamps were dominated by a now extinct order of cone-bearing gymnosperms known as the Cordaitales. The other dominant plants were tree and seed ferns. All three groups of plants – the cordaitean trees and the two types of fern – had root structures that demonstrate that they inhabited a swampy region but not necessarily a salt-water one - as we have seen, fresh-water swamp plants can have many of the same adaptations as mangroves.

What is intriguing is the virtual absence of the lycopods from this region. Were they unable to

The mangrove apple has the capacity to develop a new tree from an old branch: if a branch is so heavy that it can no longer support itself and sags into the mud, roots can develop at the contact point and a new stem will grow.

develop here because of salt water? The presence of the ferns could indicate several things: they could identify the landward edge of the mangal or, if they are later in geological time, a succession from a brackish to a fresh-water habitat, or if there is a continuous succession of the cordaitean trees and the ferns, an alternation of habitats related to changing salt- and fresh-water conditions. The evidence is not conclusive and there are others areas of investigation that need to be explored to see what independent criteria there are for suggesting this swamp system was indeed inundated by salt water.

Photosynthesis

A detailed discussion of mangrove photosynthesis is beyond the scope of this book. Suffice it to say that research into the subject has produced conflicting results. There are two different biochemical processes whereby plants can photosynthesise (synthesise carbohydrates from carbon dioxide and water using sunlight). The majority of plants are termed C3 plants: they use the pentose phosphate pathway. A C4 plant employs the dicarboxylic acid pathway. This method requires lower concentrations of carbon dioxide. Research has produced conflicting results as to which pathway is incorporated by mangroves (let alone whether they all incorporate the same pathway).

Certain mangroves (for instance some of the Rhizophoraceae) have leaves orientated vertically. One might have thought that a leaf that is square on to the sun would absorb more light and so photosynthesise more efficiently. The trouble is that in the tropics a leaf that is square on to the sun can overheat: the temperature of mangrove leaves artificially held facing the sun in northern Queensland, reached 45° C. Photosynthesis in leaves is thought to be optimised at 30° C so the vertical orientation is presumably a device to keep the leaf cool. Indeed measurement of the temperature of such vertically orientated leaves showed them to be within a few degrees of the atmospheric temperature.

Plants have tiny pores, called stomata, on their leaves (most numerous on the underside), which are crucial for gas exchange between the inside of the plant and the external atmosphere. The carbon dioxide required for photosynthesis is taken in by the stomata. Research on certain species in an Indian mangal showed that they tended to have their stomata open until about 10 or 11 a.m. and that they then closed them and kept them closed until the next morning. This indicates that photosynthesis in this particular mangal occurs during the cooler

In Buccoo Swamp, Trinidad, red mangroves co-exist with Nymphia amazonum *lilies. The presence of the lilies suggests there is little salt water in this part of the swamp.*

part of the morning. The closing of the stomata thereafter prevents excessive water loss during the heat of the day and limits further photosynthesis by halting carbon dioxide intake. Unlike some succulent plants, mangroves do not appear to be able to store carbon dioxide in their tissues overnight, nor do they again photosynthesise in the cooler late afternoon hours. Vertically orientated leaves, like those of the yellow mangrove, have stomata located only on the underside and they are surrounded by tiny hair-like projections that further reduce evaporation from the leaf undersurface. These adaptations demonstrate the limited window for photosynthetic activity available to mangroves.

An exception to this general pattern are the pencil-rooted mangroves. The genus is thought to be the most transpirationally active of all mangroves. The grey mangrove was studied by researchers and shown to reach a peak of transpiration in mid-morning. Thereafter the stomata slowly closed. However a second peak was recorded in the late afternoon. It was hypothesised that the second stomatal opening was related to the incoming tide and the resaturation of the substrate.

Salt tolerance

A noticeable feature of many mangrove leaves is their waxy texture. Thick cuticles on both leaf surfaces are an adaptation to limit water loss through evaporation from within the leaf. In other words mangrove plants must guard against dehydration because of the salt water in which their roots are often bathed. The number of water molecules within the plant for a given volume is greater than that of the surrounding sea water because the sea water has salt dissolved in it. Therefore the rate of diffusion of water molecules from the plant root membranes into the surrounding salt water is greater than that from the sea water into the plant. The plant thus risks dehydration through water loss. Taking in salt water would eliminate the diffusion of water from the plant to the sea but dissolved salt is toxic to most plants. And furthermore flooding the internal tissues with dissolved salt in effect reintroduces the problem of dehydration because the cells need fresh water: in other words they need to take in yet more water to reduce the concentration of salt.

Mangroves employ several techniques to overcome these problems. First, they have the ability to filter much of the salt out of sea water even as it enters the root membranes. The process has to overcome the tendency of the plant to lose water to its surroundings because of the higher osmotic potential of sea water. The mechanism is thought to be a passive one whereby the transpiration stream of water through the plant (the passage of water from the roots through the stem and branches and out through the stomata in the leaves by evaporation) draws salt water through the root membranes and across some sort of ultra-filtration barrier. The effectiveness of this process in a particular species is reflected by whether or not the species also contains salt glands. For example the club and the river mangroves both manage to prevent about 90 per cent of the dissolved salt in the sea water from entering their xylem (the vascular tubes that carry water and minerals to the rest of the plant). Both have salt-excreting glands in their leaves. On the other hand mangroves such as the prop-rooted mangroves and the sandy mangrove, which do not have salt-excreting glands, reduce the salt concentration in their xylem fluid to less than 1 per cent of that of salt water.

It used to be thought that mangroves were capable of actual pumping excess salt into certain 'sacrificial' leaves that were then discarded, but this is no longer thought to be the case. The salt concentration of leaves both young and old is constant. Mangrove leaves tend to have a high osmotic potential: they draw water into their cells. This is presumably important as a contributory method (combined with transpiration) of drawing water up through the plant and in through the root membranes. As a mangrove leaf ages it becomes progressively more succulent as a response to more salt being deposited in the tissues of the leaf: it draws in yet more water. Furthermore there appears to be a correlation between the thickness of the leaves of the red mangrove and the degree of salinity of the habitat in which it finds itself. The turgidity of the leaves is the osmotic result of the greater amount of salt in each leaf. It would be interesting to know whether salt concentration in the leaves of a single mangrove species is constant irrespective of the salinity of the habitat in which the trees are located. If so then two questions arise: first whether the mechanism of salt retention in the leaves and the corresponding succulence is a limiting factor in the range of salinity tolerated; and secondly what the process is that allows the water-storage tissues of the leaf to grow to the volume appropriate for the regulation of the salt concentration. Is this latter due to the passive 'elasticity' of the water-storage cells or to a homeostatic mechanism?

Mangroves, like other halophytes, have developed a degree of salt tolerance. We have seen how the viviparous seedling of the Rhizophoraceae is prepared for its salt water habitat by being progressively exposed to higher concentrations of salt passed into it from the mother plant. While this explains the method of preparation, it is a further question what the internal processes are within the seedling (and the adult plant) whereby this tolerance is achieved. A variety of complex mechanisms have been suggested to account for salt tolerance at the cellular level. Different mechanisms may be used in different types of mangrove (and different parts of the same plant) but they are likely to include osmoregulation and the active concentration of ions to offset diffusion across membranes. Indeed a general feature of salt-tolerant plants is the ability to tolerate unusually high levels of solutes in their cells, thus reducing the tendency of diffusion from within the cell to an external solute-rich medium. Stomatal closing has already been mentioned as a crucial method of limiting evaporation but the succulence of the older mangrove leaves presumably also fulfils a storage rôle even if the water is somewhat salty.

As one would expect with halophytes the rate of transpiration, which is controlled by the opening and closing of the stomata, is reduced as the external salt concentration of the habitat increases. At the level of the individual tree it is reasonable to assume that the transpiration rate is a flexible response to the daily and seasonal variations of climate and other features pertinent to the tree's survival.

Both the red and the grey mangrove have been shown to grow best in brackish water and adequately, but less well in fresh water. This shows that while they can tolerate salt water of sea water concentration (about 36 ‰ of NaCl), they prefer more dilute conditions. Given their capacity to survive in fresh water but their inability to compete with fresh-water swamp plants, one might (at the risk of playing at semantics) describe them as being obligated to their rôle as facultative halophytes.

Plants of hypersaline sand flats

There is an upper limit of salt concentration for mangroves; this is about 90 ‰ of NaCl. Higher concentrations of salt can be tolerated by salt marsh herbs that are often seen on salt flats immediately behind the mangal.

These salt pan plants have even more succulent leaves than mangroves and many are prostrate forbs. Where mangroves abut a salt pan there are often such plants immediately behind the mangal but perhaps extending only a few metres across the salt flat. Examples from hypersaline sand flats abutting Australian mangal include *Halosarcia halocnemoides* and *Tecticornia australasica*.

Mangroves can only tolerate a certain amount of salt. In arid regions such as the Northern Territory, hypersaline sand flats often occur on the landward edge of mangroves. Here, various species of mangroves culminating in the sandy mangrove, abut a hypersaline sand flat. The prostrate, brownish plants are non-mangrove salt pan plants that penetrate a little further onto the salt pan though its excessive salt soon halts them as well.

A widespread example from the New World would be the southern glasswort (*Salicornia perennis*). Beyond these highly salt-tolerant plants a salt desert appears where no plants can survive.

The replacement at the hypersaline sand flat of mangrove trees by prostrate, succulent herbs is a dramatic transition. Exactly why this increased concentration of salt defeats mangroves is open to conjecture. The plants which replace them are again presumably facultatively limited to the hypersaline region (they have been shown to grow better in lower salt concentrations).

Work on salt-marsh plants of temperate latitudes shows that they have similar adaptations to mangroves for dealing with the salty environment. The osmotic pressure in the leaves is often higher than in the rest of the plant due to the concentration of salt there. Salt glands are found in some species while others develop succulent water-storage areas in their tissue, which can be discarded at the onset of cooler or wetter weather. Other salt-marsh plants lack such structures and processes and simply shed their (salt-rich) leaves at the end of the season.

Succession

The notion of succession presupposes that there will be a change in the species occupying a specific habitat over time. An ecosystem is the interactive relationship between a community of organisms and its physical setting. The older the ecosystem the greater its stability: the rate of change – of replacement of organisms – gradually slows as the ecosystem becomes more stable, finally reaching what is termed the climax community. It is thought that the process can take hundreds of years.

The notion of succession leading to a climax community has been a dominant one amongst biological scientists for many years. Biologists have constantly tried to extend the notion of evolution to larger and larger units; the notion of succession as applied to the mangal is a case in point. One might ask, however, to what the notion refers if indeed the species composition as well as the physical setting are both changed during the course of the succession. And the notion of the climax community – the stable end-point – has itself been challenged as we will shortly see.

Nevertheless there can be little doubt that the species composition of a habitat changes over time and that those changes can be due to the effects on the physical setting of the organisms inhabiting it or of outside forces. Let us consider first what might be termed a successionist account of how the mangal becomes established and then discuss how it has been suggested it evolves into other ecosystems.

The origin of a mangrove swamp

A New World mangal that has the four common mangrove species (the red, black and white mangroves and the buttonwood) will have them in the same successive order: red mangroves at the seaward edge with black mangroves a little behind them followed by white mangroves and buttonwoods. Although one will find variations in the degree of overlap between species one is never going to find a mangal with buttonwoods at the seaward edge, followed by white mangroves and finally black and red mangroves flourishing in the higher, less frequently inundated areas on the landward edge.

Theories conflict about how a mangrove swamp actually becomes established. Perhaps the most vivid is that of the Aboriginals of the northern Australian coast. According to their traditional version, the prop-rooted mangroves simply walked ashore from the sea to take up position along the coast.

Hypocotyls develop on a stilt-rooted mangrove.

This stilt-rooted mangrove hypocotyl has been carried to a suitable substrate. It has put down roots, righted itself and just begun to put up its first pair of leaves.

There is what might be called a traditional scientific account, in which the process is also initiated by the prop-rooted mangroves. If one considers a New World mangal then the story begins with the red mangrove. There are two features central to the story: first, the fact that the red mangrove produces a seedling capable of surviving in sea water and secondly the fact that the red mangrove can survive in salt water of sea water concentration: in other words at the seaward edge of the mangal.

The story begins with a red mangrove tree developing its seedlings until they are about 30 cm in length. They are then dropped from the tree. If the tide is in, they splash into the water, where they float. If it is out they land in the mud at the base of the tree. Here a popular myth can be discounted: it is commonly believed that the seedlings are designed to fall like spears into the mud, and then develop underneath the mother tree. Such a process would make little sense: a seedling would have little hope of survival if it was doomed to develop under the canopy of a healthy, mature tree of the same species. Furthermore all those seedlings that fell when the tide was in would be wasted. The dispersal rate would be pitiable; the time needed for mangroves to colonise new habitats gigantic. In fact it has been shown that very few red mangrove seedlings that fall to the mud beneath the parent tree succeed in developing there. Because they float, the tide lifts them off the mud and away.

A few days later the pointed end (containing the primordial root system) of the seedling takes in water. The opposite end (containing the primordial leaf system) is covered in a water-repellent cuticle. Air spaces inside make this end lighter than water and the effect is that the seedling now floats with the pointed end down. It has been estimated that it can float in this way for up to a year.

Suppose a lucky seedling floating in this way eventually drifts inside a reef and is carried to the sheltered lagoon. A new sandbank has formed just below the

This shallow area has several red mangrove plants trying to become established. The stems grow as high as possible to maximise the amount of time in which photosynthesis is possible.

surface and just offshore thanks to a recent storm. The seedling drifts until it encounters the sandy bottom. Perhaps the gentle current working away at it pushes the water-logged end containing the roots into the sand. Or perhaps the outgoing tide leaves it lying flat on the sand; either way, the roots grow downwards and the developing plant rights itself and begins to grow. If the plant encounters the sand bank at a lower level (thanks to the height of the tide that brought it there) then it has greater problems than if it had taken root higher up the sand bank as it appears that daily submergence of the seedling by more than 20 cm of water has an inhibitory effect on germination.

Rooted in the soil and vertical, the next action of the seedling is dramatic: it uses every ounce of energy it has to grow its stem upwards, as high above the surface level as it can reach. A pair of small, fragile leaves unfurl at the top of the stem so that photosynthesis can begin. By the end of the first year the plant is a metre tall but desperately fragile. However from then on it begins to produce the prop roots that are such a familiar feature of the

Mangrove oysters (Crassostrea rhizophorae) have attached themselves to the prop roots of this young red mangrove tree in the Bahamas. A micro-habitat will become established on the roots.

species. It grows ever more stable. A crown of leaves is produced. Meanwhile other red mangrove seedlings have drifted to this sand bank and taken root. Within a few more years a bank of saplings, each with its numerous prop roots, is established. The prop roots slow the current. Sediment run-off from the shore is slowed and particles of mud sink around the mangroves. Some of their own leaves are similarly collected at the base of the young trees. They attract bacteria and fungi that start to break down the leaves into detritus. The transformation of the substrate from relatively sterile sand into a muddy, detritus-rich, anaerobic medium is underway.

This new substrate becomes home to all sorts of algae and animals as well. The prop roots of the red mangroves that are beneath the limit of the low tide are soon covered in a rich tapestry of algae, filter-feeding sponges and ascidians, as well as hydroids. Oysters and mussels take up station wherever their larvae can attach. Nutrients carried out of the young mangal enrich the surrounding waters and sea grasses begin to appear in the sand beyond the young trees. Numerous fishes – both herbivores and carnivores – become established in this new biome, as do countless varieties of worm, crustacean and mollusc.

Years go by. The sediment-trapping ability of the red mangroves means that the substrate is a little higher than before. This allows black mangroves to become established. Continued sediment-trapping over the years produces yet higher levels of mangrove mud where the other two mangrove species common to this corner of the New World, the white mangrove and the buttonwood, now appear.

This young red mangrove plant is a few years old. Note the pneumatophores of a black mangrove that is a little higher up the intertidal domain. According to the successionist model, the presence of the black mangrove spells doom for the red mangrove.

*A great egret (*Egretta alba*) hunts in stately slow motion at the brackish landward edge of Caroni Swamp in Trinidad. The extensive mangrove forest slows the drainage capacity of the area and pools of fairly fresh water result, in which needle grass thrives. The large (out of focus) trees in the background are black mangroves benefitting from the fresh water input to the area.*

Evolution

In the same way that the successionist interpretation has been applied to the mangal it has also been used to try to show how a mangrove ecosystem is doomed to evolve into a terrestrial one.

A common sight at the landward edge of a New World mangrove is a low-lying area of trapped fresh water that is dominated by sawgrass (*Mariscus* sp.). Such an area could have formed because the sediment-trapping of the mangal and the saturation of the soil vastly reduces the drainage capacity of this inland area. Fresh-water run-off tends to pool in low-lying areas. Black mangrove trees thrive at the interface of brackish conditions and are common at the mangal edge of such settings as are white mangroves and buttonwoods in slightly more elevated places. Shrubby, dwarfed red mangroves can occur amongst the sawgrass (presumably surviving but not thriving in the predominantly fresh-water conditions). This type of habitat is often rich in birds which leave the shelter of the mangroves to feed. An example of such an area is the marshy back edge of the Caroni Swamp in Trinidad.

The successionist explanation for such an area is that pioneer red mangroves give rise to mature red mangroves. Black mangroves appear and become dominant. Sawgrass appears. Red mangroves invade the sawgrass to a limited extent, and buttonwoods appear in the drier, more elevated places. Eventually the drying, elevating ground rises above the highest spring-tide level and a terrestrial forest appears. Meanwhile at the seaward edge the pioneering red mangroves are constantly having to re-establish themselves as the other mangrove species usurp the ever-rising substrate habitat. And so, according to this theory, do succession and progression occur: succession of the species composition of the mangal and progression in the indefatigable advance of the red mangrove seawards as it claims ever more land from the sea.

A group of scarlet ibises (Eudocimus ruber) feed in the brackish back edge of Caroni Swamp. Dwarf red mangroves are common in this region.

This model of mangrove succession is still a popular one, although we have already seen that mangrove peat cannot build up land, because it disintegrates when exposed to the atmosphere, and that the red mangrove does not reclaim land from the sea; it colonises regions of natural sediment accretion and can help trap sediment to be sure, but a mangrove swamp is no defence against coastal erosion. Furthermore the theory implies that the non-seaward edge mangroves can only become established in substrate originally occupied by (and prepared for them by?) the red mangrove. This is false. Most mangroves can tolerate a range of substrate types: taking root in an appropriate intertidal location is the criterion for development.

Any attempt to explain how mangroves develop is likely to have to take account of other factors, such as forest fires and severe weather conditions. Storms, hurricanes and cyclones can destroy mangrove forests in a variety of ways - most obviously they can blow the trees down. Excessive wave action can also sweep away the sediment into which the mangroves are

The cedar mangrove is readily identified by its peeling bark. This is a deciduous mangrove and therefore more likely to survive having its leaves torn off by a cyclone than its evergreen neighbours.

anchored or deposit new sediment in the mangal above the height of the aerial roots of certain species so that the trees suffocate. Strong winds can strip the leaves from the trees. Evergreen mangroves will often perish, although deciduous mangroves such as the cannonball mangrove are more likely to recover. However even amongst the evergreens some are better able to survive storms than others. The black mangrove is more vulnerable than the red mangrove. The effect of severe storm damage can completely alter a mangal. Sometimes mangrove ferns become so dominant that no other species can establish itself.

Zonation

Certain mangrove plants are generally found in fairly exclusive bands that parallel the shoreline. We have already discussed New World examples. In the Indo-Pacific region, the mangrove apple is found at the seaward edge of the mangal and the sandy mangrove at the landward edge.

Not only will the conditions of a particular location (such as the tide and drainage characteristics of the soil) affect the state and abundance of a species in a particular mangal. There are also factors like climate (seasonal variations in temperature, evaporation and rainfall) and the geographical dispersal pattern of the species in question. For example in arid regions such as the Northern Territory of Australia, the back of the mangal is often bordered by hypersaline sand flats. The edge of the hypersaline sand flat will probably be fringed by a sandy mangrove zone.

The question we are primarily interested in here is why the mangroves in a particular mangal have the zonation that they do. One might expect that features of zonation in specific mangals could be used to explain zonation in general, but as we shall see, every attempt to do so can be challenged with an awkward counter-example. Although zonation is often striking, its explanation has proved to be elusive and may be dependent on a complex combination of variables.

For zonation to be clearly visible it is necessary for there to be enough space in the habitat offering the same range of conditions for numerous trees of the same species to become established. In other words where the mangal is cut through by a mass of creeks and streams or the substrate composition is irregular a patchwork of mangroves rather than clear zonation is likely to result.

Zonation is often quite noticeable to anyone working their way transversely through a well-developed mangal. In a Northern Territory mangal one might enter at the landward edge through a belt of yellow mangroves perhaps 40 m in depth. As one proceeds, knee-rooted mangroves become more common until one is in a zone dominated almost exclusively by knee-rooted mangroves. Where a creek runs through the area its banks will be bordered by stilt-rooted mangroves. After perhaps 50 m of knee-rooted mangroves one enters a zone of stilt-

*The lemon-breasted flycatcher (*Microeca flavigaster) *is one of the few birds of Northern Territory mangals that does not immediately dash off when you try to photograph it. It has a beautiful song and feeds on small insects in flight.*

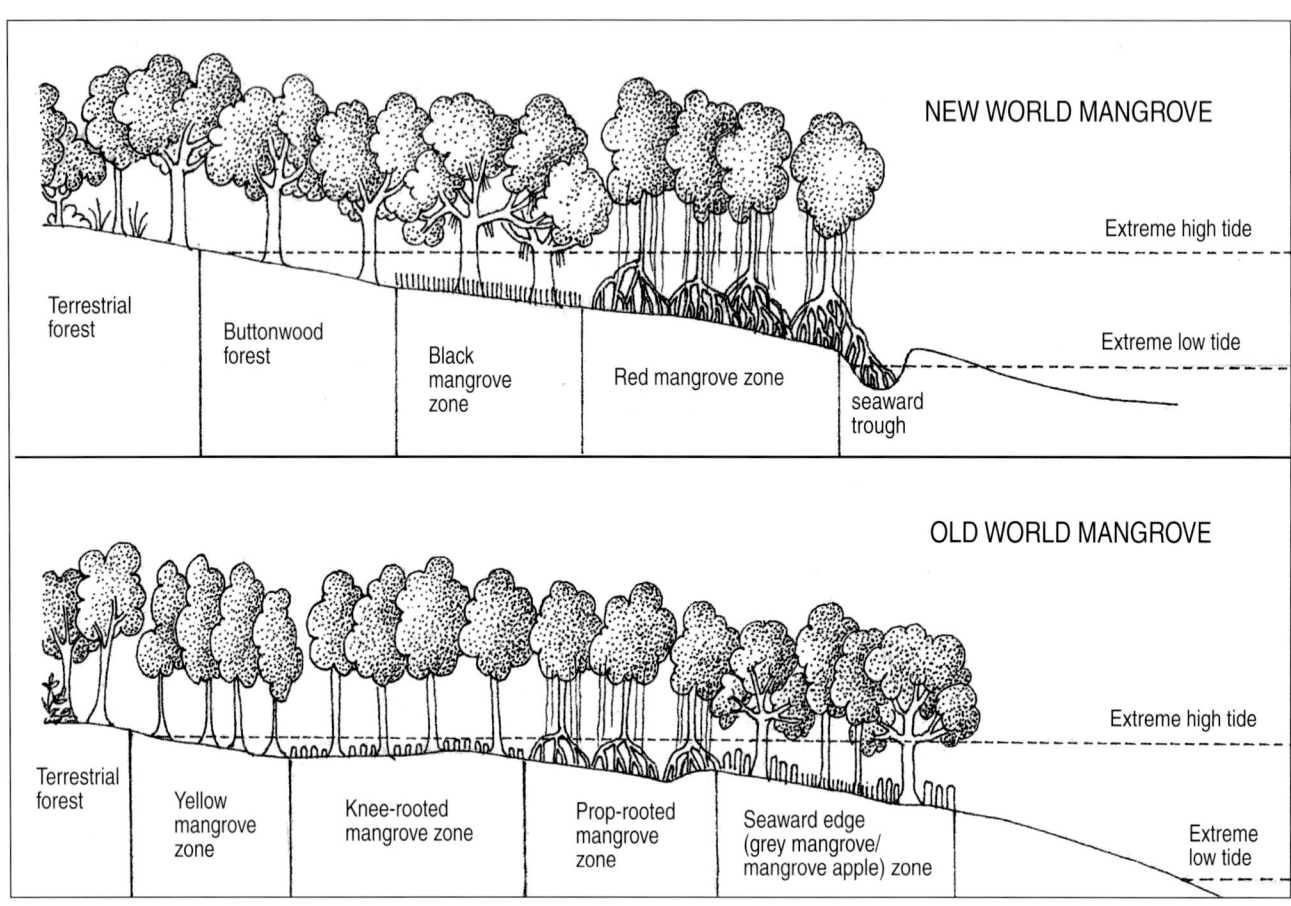

Figure 10. Examples of common zonation patterns from a New World mangal and an Old World mangal.

rooted mangroves perhaps another 50 m deep. Finally at the seaward edge one enters the more spacious mangrove apple zone.

The zonation diagrams opposite, although artificially simplistic, demonstrate typical zonation patterns in a New World and an Indo-Pacific mangal.

The most obvious feature of zonation is its apparent relationship to the tide. But what are the relevant components of the tide that cause zonation?

As a rising tide meets fresh water in a river mouth one encounters a salinity gradient: gradually the concentration of salt in the water decreases until none is present. The salinity gradient is the most obvious feature for attempting to explain mangrove zonation: the salinity of the environment determines the species present. However work done on the salinity of the corresponding mangrove zones in a Jamaican and a Floridian mangal showed a wide difference in salinity between the corresponding zones. It would be interesting to know whether any such salinity comparison has been done in the eastern region, for example comparing the Northern Territory mangal (an arid region with high salinity build-up due to evaporation) with the northern Queensland mangal (subject to heavy, if seasonal, rainfall).

The seaward zones of neighbouring eastern region mangals can be populated by different monospecific stands - a seaward edge of mangrove apples could neighbour a seaward edge of the grey mangrove. The salinity and inundation are likely to be almost identical in both sites, so the explanation may lie in a combination of biological factors and chance: interspecific competition or exclusion mechanisms initiated where pioneering propagules happen to become established.

Earlier discussions on ecosystems in general tended to assume that the more established an ecosystem, the less influenced it is by external forces. While this view may be useful, it should be asked how useful it is in each case. It is natural to think that zonation is a reflection of some such developing stability within a mangal, and as I have said, tolerance to the concentration of dissolved salt is often thought to be the first clue to the cause of zonation. However actual measurements of the salt concentration in the substrate of a given mangal over an extended period may well show radical changes that are independent of the zones. The factors affecting salinity are numerous: fresh-water input from rainfall, underground seepage and terrestrial run-off can vary enormously from year to year. The characteristics of the soil itself can change markedly as a result of inorganic processes such as sediment deposition; meanwhile the organic processes in the soil that are a response to the litter production of the mangroves (itself tied in to numerous variable influences) can alter the salt-carrying capacity of the soil and its evaporation characteristics – and the higher the rate of evaporation, the greater the salt build-up. In other words the range of factors to be considered may be enormous and an attempt to explain zonation in the simple terms of salt concentration is doomed. Thus salt concentration has been proposed as a 'competition eliminator': within its extremes it eliminates certain plants. The problem is to explain the apparently stable process of zonation in the light of numerous variables. Scientists have drawn up tables which correlate species of mangrove and their distribution through the mangal to tidal influence. Table 5 is such a table for mangrove communities in Papua New Guinea.

The trouble with elegant correlations between mangrove species location, zonation and tidal inundation classes is that several species of mangrove can turn up almost anywhere in the mangal. Laboratory experiments and field observations have both shown that some of the mangroves that are found naturally on the seaward edge thrive best in brackish water, suggesting that there is some other reason why they do not regularly become established in

Table 5. *Inundation Classes of Mangrove Species in tidal forests of Papua New Guinea.* From Womersley (1983).

Class 1
Inundated (at least partially) by all high tides:
    Avicennia spp.   *Sonneratia alba*

Class 2
Inundated by medium high tides:
Avicennia spp.	*Rhizophora apiculata*
Sonneratia alba	*Rhizophora stylosa*
Rhizophora mucronata	*Bruguiera parviflora*

Class 3
Inundated by normal high tides, although not necessarily by salt water:
Rhizophora mucronata	*Bruguiera* spp. (but not *B. parviflora*)
Rhizophora apiculata	*Xylocarpus granatum*
Rhizophora stylosa	*Xylocarpus moluccensis*
Ceriops tagal	

Class 4
Inundated only by spring tides:
Bruguiera spp.	*Xylocarpus mollucensis*
Xylocarpus granatum	*Lumnitzera* spp.

Class 5
Only rarely inundated by exceptional or equinoctial tides:
Sonneratia caseolaris	*Heritiera littoralis*
Lumnitzera spp.	*Hibiscus tiliaceus*

more landward zones. The picture becomes even more confused when one realises that some seaward edge species such as the grey mangrove occur throughout the mangal and actually tend to form much more impressive monospecific stands in areas other than the seaward edge. I remember seeing a mighty grey mangrove forest bordering a tributary of the Adelaide River in the Northern Territory: it was a magnificent expanse of grey mangroves nowhere near any seaward edge. Moreover, it is by no means unusual to find grey mangrove trees at the very back of Northern Territory mangals on the very edge of salt pans! In fact recent work on Papuan mangals has shown that many species can occur in greatest numbers in the two extremities of the same mangal: the seaward and landward edges. This hardly confirms the idea that tidal influence is the central feature for explaining zonation, although it may be that the range of a species through a mangal is limited when certain more efficient competitors are present.

It has been suggested that if salinity is not the tidal factor determining zonation then perhaps it is the amount of time a particular zone remains unflooded. Thus in a Caribbean mangal it was found that the red mangrove zone had 0-10 unflooded days annually, the black mangrove zone 10-110 and the white mangrove zone up to 157. One would need to know how closely these figures correspond with the zonation of other similarly comprised mangals and also at what stage of a plant's development this feature – if valid – is relevant. It has been suggested that it is seedling germination that is the critical factor. But experiments have shown that seedlings from the higher zones can germinate at the seaward edge, although whether they can develop into healthy adult trees is another matter. Certainly there is an experiment begging to be done: a comparitive study of rates of development by the different mangrove species when subjected to different amounts of time during which the substrate is unflooded.

Another theory attempted a physical explanation: that the size of the propagule determines the zone in which a species would tend to develop. Thus those with large, heavy propagules would tend to occur at the seaward edge because the tide transports them a shorter distance and vice versa. However there is little evidence to support this theory. It is true that the stilt-rooted mangrove has relatively large propagules (up to 40 cms in length) and is found in the lower intertidal range. However its close relative *Rhizophora mucronata* has the largest of all mangrove propagules (up to 60 cm in length),

The fruits of a mangrove apple give the tree its common name. This is a seaward-edge mangrove with relatively small propagules, which disproves the theory that propagule size determines zonation since one would expect mangroves with much larger propagules to occur lower down the intertidal range than the mangrove apple.

and where it occurs with the stilt-rooted mangrove it tends to be higher up the mangal than its cousin. Similarly the yellow mangrove, or at least one variety of it, has propagules that can reach a relatively large size - 25 cm - and occurs at the landward edge of the mangal, whereas at the seaward edge one finds the mangrove apple whose propagule is a modest 4 cm.

The successional model attempted to explain zonation in terms of species replacement due to substrate elevation. It presumed that mangroves are transitional to a climax community of terrestrial plants – the only plants capable of replacing themselves over a relatively long period of time. The land-building theory of mangroves has been knocked down at least in so far as the mangal is supposed to evolve into a forest community above the intertidal height.

What is intuitively powerful about the traditional successional model is the idea that zonation reflects the process whereby the mangal evolves into an ever-more stable community. A factor that may be relevant to a discussion of zonation is the state of the shoreline on which a particular mangal is established. A coastline can accrete or erode at any rate from negligible to rapid; it can be in a relatively stable state where the rate of erosion is balanced by sediment deposition, or it can even be in a state of flux between accretion and erosion. The rate at which any or all of these processes occur could be relevant to zonation. In sheltered areas with stable coastlines, pioneer mangroves, thanks to their sediment-trapping and peat-depositing capacity, can increase the intertidal area available for colonisation by new mangroves. It would be interesting to know whether the conditions reflected by the zonation patterns in such a mangal are equivalent to those of a mangal on an accreting shore.

The establishment of a monospecific stand will shift competition from between species to within one species and heighten the capacity of that species to become ever better adapted to the numerous local conditions of that zone – but at the expense of making it less able to compete with the finely-tuned dominants of other zones. In other words it may be not so much that zones determine mangroves but rather that, within certain parameters, mangroves determine zones. On this model one could say that zonation is indeed succession to greater stability. The question then is whether we can think of a zone within the mangal as representing a climax community.

What would the criteria be for such a climax community zone? One would expect its biomass – the total mass of the organisms – to be higher than anywhere else in the ecosystem. The organisms of the climax community would be larger than those of other parts of the ecosystem. One would also expect energy transfer to be at its most efficient, and the area with the smallest temperature fluctuations is likely to be the most stable. Furthermore, the climax area should be least affected by physical effects, and the organisms there should have a high capacity for replacement.

This impressive area of knee-rooted and prop-rooted mangroves lies within a Papuan mangal. Note the extensive understorey of knee-rooted mangrove seedlings. The ability of the knee-rooted mangrove seedling to develop in shade may help to explain how knee-rooted mangroves form zones.

Research on a mangal at Hood lagoon, Papua New Guinea, has investigated just this question and found several correlations in one particular zone with the classical model of a climax community. The mangal was some 1000 m deep from its landward to its seaward edge. The knee-rooted mangrove zone consisted mostly of the large-leafed orange mangrove, but also *Bruguiera cylindrica*. The density of woody tissue and the size of the trees in this zone were greater than outside it. Measurements of the area of the leaves compared to the biomass of the trees suggested that they were more efficient at primary production than others in the mangal. The temperature range in this zone was less than that for other zones. Moreover, because the knee-rooted mangrove zone is in the centre of the mangal it is furthest from both the seaward and the landward edges and therefore most sheltered, and seedling production was shown to be very high.

It was also noted that there was a cleared area within the zone - the remains of an abandoned village - which had been colonised by four other mangrove species that had singularly failed to establish themselves in the undisturbed knee-rooted mangrove zone. This suggests competitive exclusion by the knee-rooted trees. The virtually impenetrable tapestry of knee roots that covers the ground in the knee-rooted mangrove zone would prevent most new seedlings from taking root. Furthermore while the seedlings of the large-leafed orange mangrove can develop in shade (below areas of dense canopy in the centre of the mangal) most seedlings cannot and would be unlikely to survive in the deep shade of the knee-rooted mangrove zone.

Given the ability of many mangroves to grow throughout the mangal this central exclusion zone may represent a climax community that excludes other species. On the other hand it may simply be the successful development of one group of trees under particularly beneficial local conditions. The time-scale in which an intertidal climax community is supposed to develop is likely (because of the inherent instability of the domain) to be considerably briefer than that of a terrestrial forest habitat and this raises the question of whether the notion of climax community should be applied at all in this dynamic setting.

The enormous size of the mangal studied in Papua raises questions about whether the knee-rooted mangrove 'climax' community here is representative of a tendency in all sufficiently extensive mangals or is an atypical case. Another feature of a climax community that one would expect to find is a greater number of animal species occupying more specialised niches and this result was not found in the Hood Lagoon mangal. This idea presupposes that the maturity of the mangal will be reflected in the diversity of its faunal population, but zoologists think that the majority of animals in the mangal are generalists and not obliged to be there. Furthermore if this is the case then animals entering the mangal from the seaward and landward edges would take the longest to colonise the centre and presumably have little need to penetrate that far until the outer zones were saturated. This may suggest that the Hood Lagoon mangal demonstrates a pseudo-climax community rather than a genuine one.

What the Hood Lagoon results do show is that in secure and spacious conditions mangroves can develop into sizeable forests of considerable stability and productivity. This demonstrates that the mangal is an autonomous biological unit and not a stressed, preparatory, short-term precursor to a terrestrial ecosystem. Mangroves must be studied according to their own time-

*A houndfish (*Tylosurus *sp.) hovers at dawn amongst the mangrove roots in a Papuan mangal.*

scale (which depends on the condition of the shoreline on which they are established) and in terms of their own domain – the limited space available in the intertidal zone. Their very remarkable productivity in this profoundly inhospitable domain should surely be a cause of wonder.

If it has been established that mangroves cannot build land then the question still remains whether there are any cases where mangroves have been seen to be successional to terrestrial forest. The answer is yes. On the western coast of Malaysia large river systems deposit heavy amounts of sediment to form a rapidly accreting shoreline. The mudbanks that result are colonised by successional bands of mangrove that flourish and are replaced by new species. Eventually, thanks to the prograding effect on the shoreline by the steady supply of terrestrially derived sediments, tropical forest replaces the mangrove community. It is tempting to think of this successional type as offering an accelerated view of mangrove succession in general, but it is more likely to be the effect of the rapidly changing habitat.

Linked to this is the question of whether mangroves can advance inland. In areas where the coastline is sinking, the sea level is rising or the natural barriers blocking low-lying land fail, the ingressing sea will kill terrestrial vegetation and provide for the possibility of mangrove establishment. In the Northern Territory of Australia, mangrove incursion has occurred for curious reasons. Some 282,870 feral buffalo were counted in 1982 roaming the floodplains. These huge animals love to wallow in the swampy conditions and they form corridors through the vegetation as well as gouging deep troughs through the swamp in which they wallow not only to feed but also to remove parasites. These channels and troughs are sufficiently extensive in the floodplain of the Mary River for sea water to have entered and killed off the terrestrial plants for considerable areas. The result is that mangroves are penetrating inland.

Root adaptations

As previously mentioned mangrove roots are embedded in an anaerobic substrate. It is also (at least in the lower intertidal areas) regularly submerged by sea water. The mangroves must find a way of getting oxygen into their roots at least when the tide is out so that the physiological activities of the root system can take place.

Perhaps the most familiar root modifications belong to the prop-rooted

These fine red mangrove trees are in Caroni Swamp; the prop roots are clearly visible. The tree on the left has produced a cascade of aerial roots from high up the tree. It is not clear what function these structures have and they may simply be due to growth occurring at points of damage (for example, from beetle attack).

mangroves. The spectacular arcing root formations of this group are commonly known (as the common name suggests) as prop or stilt roots. The structures also have the important function of stabilising the tree: immediately below the area where large prop roots are given out in abundance, the trunk of the tree is likely to be thinner than above it. This shows that the weight of the tree is being dispersed into the prop roots. Not only does the system offer greater support but by dispersing the weight over a larger area it prevents the tree sinking into the mud. Occasionally one will see a *Rhizophora* tree with a veritable cascade of roots given off not from near the base of the trunk, but from the crown of the tree. These grow down many metres and eventually become embedded in the mud at the base of the tree. It is not clear what function these structures perform and it has been suggested that they are given out at injury sites – for instance parts that have been attacked by burrowing beetles.

The other most conspicuous root adaptation likely to be encountered by a mangrove swamp explorer at low tide are pneumatophores. The two most commonly seen types are relatively easy to distinguish. The pneumatophores of the pencil-rooted mangroves tend, as the common name suggests, to be pencil-thin. They can occur in vast numbers and in considerable density around the base of the tree, thinning out further from it as the gap between the roots increases. They are given off at regular intervals (about 20 cms apart) along the root. *Avicennia* pneumatophores are usually about 20 cm high, although they can grow to perhaps a metre. *Sonneratia* pneumatophores on the other hand are altogether larger. They can have a diameter of 10 cm or more and in exceptional cases they grow to several metres in height although they are usually less than a metre. Their greater diameter gives them their name of peg roots and so we call the members of this genus the peg-rooted mangroves. In both groups the pneumatophores are vertical branches given off at regular

This large old black mangrove tree is at the back of Kilgwyn Swamp in Tobago; note the numerous pencil-thin pneumatophores. The tree is also, unusually, giving off numerous aerial root structures possibly as the result of insect damage.

intervals from the horizontal meanderings of the submerged root.

In the *Bruguiera* genus characteristic knee roots are developed and so the group is sometimes known as the knee-rooted mangroves. These, as the name suggests, look like knobbly knees. The knee roots are formed by the upward growth and folding over of the main root; secondary roots are given off at the site of the knee roots.

The cedar mangrove gives off an impressive display of snorkel-sized

Left: The knee roots of the yellow-flowered orange mangrove; note how secondary roots originate in the structure of the knee root itself.

Right: The snorkel-sized pneumatophores of cedar mangroves dominate this scene. A cedar mangrove is visible off-centre (it has the peeling bark).

Schultz's mangrove produces the most comical-looking pneumatophores; they are simultaneously obscene and absurd.

pneumatophores which are the result of localised development from the upper surface of the submerged root. Its cousin the cannonball mangrove has what are called ribbon roots; the dorso-ventrally flattened roots meander half-exposed through the mud.

Perhaps the prize for the strangest-looking aerating structures goes to Schultz's mangrove. The lumpy excrescences produced by this plant look simultaneously obscene and absurd. Rorschach psychologists should avoid such areas.

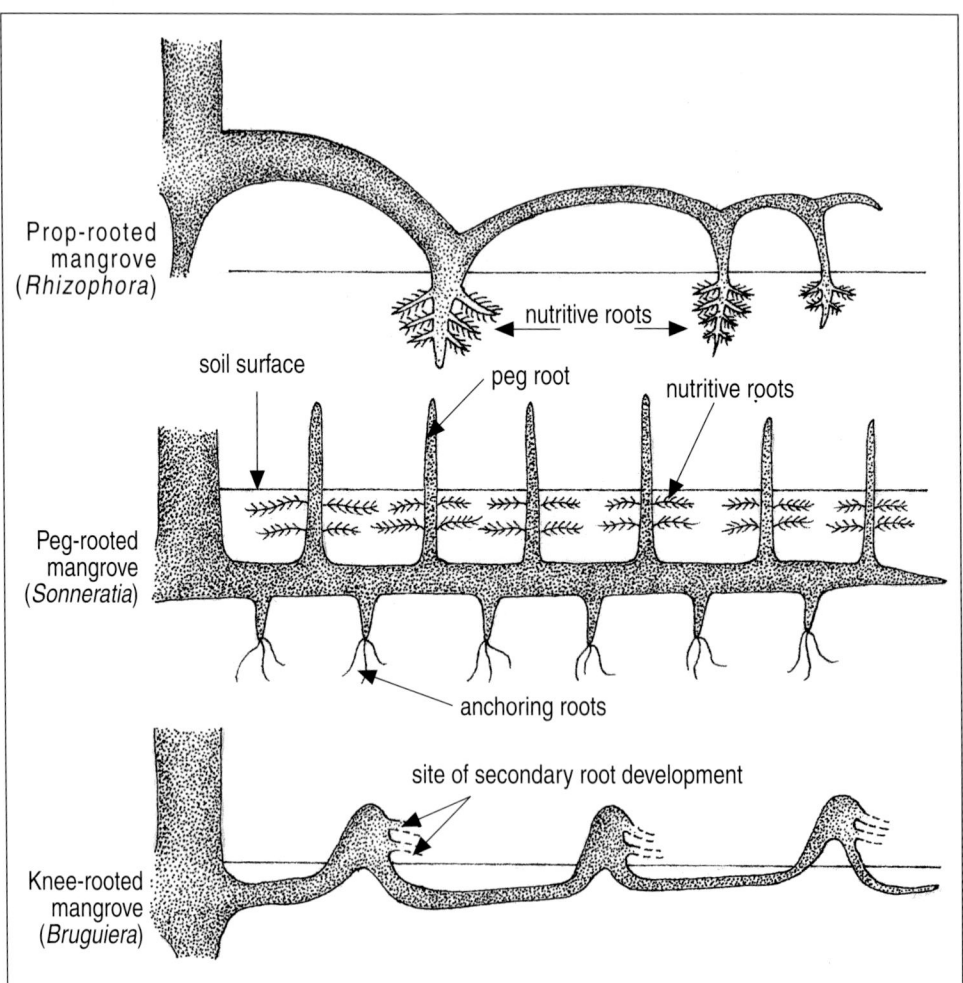

Figure 11. Diagram of various mangrove root systems.

These are perhaps the most commonly encountered types of aerating mechanism. A closer examination of pneumatophore structure reveals the presence of lenticels – small pore-like openings designed for gas diffusion, and the internal structure of the root includes extensive gas passages that lead to the stem.

A study of the exposed portion of the roots of a mangrove is often a good start for its identification, but not all mangroves give off such structures. The river mangrove, despite being tolerant of a wide range of salinity and habitat does not produce pneumatophores, and the white mangrove only occasionally does so. It is not unusual to see the roots of the river mangrove simply breaking through the mud surface and thus being exposed to air. However this may be caused by the removal of mud due to local weather conditions rather than being an aerating strategy. Other non-pneumatophore producers include the milky mangroves, the myrtle mangrove, the mangrove palm and the club mangrove, although the club mangrove may have an alternative (if less conspicuous) aerating structure: the base of the stem is expanded and spongy. When one remembers that very few mangrove associates can produce root-aerating mechanisms yet survive in the mangal one has to ask whether the aeration hypothesis is in fact the whole story or whether there are more subtle processes at work in these plants that are yet to be discovered.

The root systems of mangroves have the basic functions common to all roots: structural support, absorption and transport. Anchorage is provided by the main root system meandering beneath the substrate surface. In fact the roots do not penetrate far beneath the mud, presumably because the anoxic environment halts root growth downwards. So shallow is the root penetration (often considerably less than a metre) that mangroves can be described as floating on a sea of mud. The combination of soil erosion and strong winds can easily cause trees to collapse. It is by no means unusual to see the main root

Cyclone Tracy hit Ludmilla Creek, just outside the city of Darwin, at the end of 1974. Much of the mud was washed away. The roots of this mangrove apple in the creek have been exposed and their structure is clearly visible. Nevertheless the tree has managed to give off secondary structures from the original roots which have repenetrated the mud so that the tree can survive.

branches exposed by sediment removal, hence the requirement for a sheltered location. In some instances a mangrove can develop a lower level of roots when the first level is exposed and thus re-anchor itself. If it is subjected to regular sediment deposition then, provided the pneumatophores are not submerged, it can produce a higher level of roots. This process has been studied in the mangrove apple and has led some scientists to suggest that the most significant purpose of the aerating mechanisms is not gas exchange but rather to allow the mangrove to adapt to changes in substrate level.

The absorption of nutrients is performed by the finely branched pinnate terminations of the root system which, not surprisingly, are located in the most nutritively rich and most oxygenated layer of the substrate: immediately below the surface.

As we have observed there are extensive hollow passages in the roots of mangrove plants that lead to the stem. Given the enormous number of pneumatophores produced by mangroves such as those of the pencil-rooted genus, it seems likely that gas exchange is not merely a matter of diffusion. Furthermore the distances involved in a large tree would suggest that an active process is required for adequate aeration.

An elegant solution has been suggested – at least for pneumatophore-bearing mangroves that are frequently inundated by the tide. When the tide is out the pneumatophores are exposed to the atmosphere. The lenticels are open, allowing air to enter the root system. When the tide comes in and the lower part of the tree is submerged the lenticels close. Meanwhile respiratory activity continues within the plant. The result is that oxygen is used up while the carbon dioxide produced, being soluble, will tend to dissolve in the fluid medium inside the plant. The result is a drop in pressure inside the plant, resulting in a slight vacuum. When the tide goes out and the pneumatophores are again exposed the lenticels open and air is literally sucked into the root system.

This theory has been supported by measurements of oxygen and carbon dioxide levels in submerged pneumatophores that show that they vary as predicted. Furthermore the fact that mangrove roots do not plunge great depths into the anoxic mud suggests that valuable as the air suction system is it is of limited effectiveness. Diffusion must suffice higher up the shore where tidal immersion is rarer and the substrate presumably holds more oxygen.

The 'vacuum hypothesis' could also be another reason why mangroves in areas of greater tidal influence produce more leaf litter (i.e. respire more), as they have a method of regularly flushing their root systems with atmospheric oxygen.

Flowers and flowering
Struggling through a mangrove swamp in Australia's Northern Territory in November 1993 I reached an area where there was a pungent, over-ripe smell in the air. My first suspicion was that I was about to stumble upon a fisherman who needed a good wash and had tried to disguise the fact by liberally dosing himself with a bargain-priced after-shave.

As I continued along the creek bank the clammy smell seemed to be irregularly dispersed: it would invade my nostrils only to fade away a few metres further on. I peered through the vegetation trying to work out the cause of the stench but could find no answer. I sank up to my knees in a soft patch of mud, and the pungent odour was all around me. I grabbed the pale, smooth branch of a grey mangrove tree and used it to pull myself out of the mud. As I did so I noticed that something was subtly different about the branch but at first I could not recognise what. Looking closer I realised that there were tiny pale orange flowers bursting from amongst the leaves. A cautious sniff of the flowers established the source of the heady aroma.

The tiny orange flowers of the grey mangrove produce an aroma out of all proportion to their modest size.

I had not visited the Northern Territory in November, the build-up to the wet season, before; my previous visits had been in June and July, during the dry season, and I had not seen many mangroves in flower. But in November the humidity is exhausting. Huge thunderstorms rumble away in the distance. When the sun breaks through gaps in the gathering clouds the temperature ranges from very, very hot to even hotter.

As I continued to work my way through the mangal I decided to look more closely at other plant species to see if they too were in flower. At the seaward edge of the creek bank was a low clump of club mangroves. At the ends of their branches were clusters of tiny white flowers. On the way back through the mangal I continued my flower-check. The yellow mangroves were long past flowering: their branches sagged under the weight of soon-to-be-released propagules. But at the landward edge the sandy mangroves were also covered in tiny white flowers. Wasps buzzed from flower to flower.

So small and inconspicuous are the flowers of most mangroves that they are easily overlooked. Small flower size and pallid coloration are usually indicative of insect pollination. (The pale coloration suggests that there are ultaviolet attraction signals in the flowers that are visible to insects.)

The three sexual states demonstrated by true mangroves are: when male and female reproductive organs are found on different plants (dioecy), when the bisexual plant has both male and female reproductive organs (flowers) although the flowers are separate (monoecy); and the hermaphrodite state when both the male and female reproductive organs are contained within the same flower. Of these three states dioecy is the rarest and monoecy occurs more frequently, but the hermaphrodite state is by far the most common. Tomlinson has calculated that at the generic level 74 per cent are hermaphrodite, 16 per cent are monoecious and 10 per cent dioecious.

The small, pale flowers of most mangrove species are pollinated by bees and wasps. This sandy mangrove is in flower and being visited by various wasps. Feeding on the flower in the foreground is the sphecid wasp Bembix littoralis.

The great advantage of sexual reproduction over asexual is in the ever new combinations of genetic material formed by the fusion of the male and female sex cells. The genetic recombinations allow for adaptation via mutation and are thus the first stage of evolution (the next being natural selection). On the other hand in asexual reproduction the offspring has the same genetic material as the parent; it has all the advantages and all the limitations of its parent.

The problem of sexual reproduction is that the male and female sex cells need to be brought together to fuse and form the zygote – the single cell that develops into the embryo of the offspring. For example in a plant that has both male and female reproductive organs there is rarely any point in the plant fertilising itself, as the result is little better than asexual reproduction. Thus in non-dioecious plants strategies tend to be employed to prevent self-fertilisation. One tactic is to ensure that the male and female sex organs mature at different times (dichogamy) so that a male sex cell (the pollen) from one plant can only fertilise the ovule of another plant. Where a plant first produces female gametes and then male gametes the plant is termed protogynous. An example from the mangal is the mangrove palm. The alternative dichogamous condition is termed protandry: here the pollen matures before the production of ovules. The pencil-rooted mangroves are protandrous.

The mechanisms of pollination in mangroves are far from fully understood. In some plants the male and female flowers are markedly different while in others the flowers are almost indistinguishable. In the woody-fruited mangroves the dioecious flowers are virtually identical in appearance. However the male flowers produce infertile ovules and the female ones produce infertile pollen. Identical nectar is found in each flower type. A cunning strategy is thought to be at work here: a visiting insect that feeds on a male flower and collects pollen will, when it goes in search of other similar flowers, be just as likely to arrive at a female flower and so fertilise it.

It has been suggested that for mangroves to establish themselves in isolated places and increase rapidly in number, self-fertilisation would be an advantage. This raises some interesting possibilities: for example could colonising mangroves have inbreeding mechanisms which are subsequently suppressed (by intraspecific competition or proximity) when the number of mangroves increases? If crowding generates dichogamy (the flowering of male and female flowers at different times) then it would be fairly easy to monitor the flowering processes of isolated monoecious or hermaphrodite mangroves that are usually considered dichogamous to see if their flowering processes are, when isolated, synchronous. For example in the protandrous hermaphrodite black mangrove the flowers, although located in groups, tend to open singly. If this is a means to prevent self-fertilisation (the vector is thought to be various species of bee) then isolated, colonising trees might have a different timing for flower opening - or, more specifically, have flowers producing pollen and flowers producing ovules open at the same time. This is a question which, to my knowledge, has yet to be investigated.

Certainly the fertilisation by pollen from other flowers of the same plant (geitonogamy) has been suggested for at least one species of mangrove that is, significantly, a pioneer species: the mangrove apple. While other mangroves that employ vectors mostly employ insects, this species (and the other members of the genus) is remarkable in that it chiefly employs bats.

The large (by mangrove standards) flowers of the mangrove apple open at the end of the day and are accompanied by a characteristic odour. A generous supply of nectar is held within the bowl formed by the outer ring of flowers. At night bats come to feed on the nectar and are thought to be the major pollinating agents for the species. Indeed at least one species of bat (*Macroglossus minimus*) is thought to be completely dependent on the nectar produced by the peg-rooted mangroves. Various birds such as honeyeaters will also readily

Ants (Formicidae) feed in a black mangrove flower. Although flowers occur in groups in this species they tend to flower singly, perhaps to prevent self-fertilisation.

feed on the nectar. I have seen brown honeyeaters (*Lichmera indistincta*) do so and they may also be responsible for pollination, although the flowers are primarily timed for nocturnal visitors: by the next morning they are disintegrating and the petals and stamens (the pollen-bearing structures) are being shed. Where bats are rare or absent it is thought that hawk moths (Sphingidae) assume the rôle of pollinators.

While Northern Territory mangrove apples tend to flower throughout the year, marked seasonality has been recorded in their flowering in Queensland : they flower only in the last three months of the year. If perennial flowering is the primitive condition and seasonal flowering developed from it, seasonal flowering is likely to be a response to the limited availability of a major pollinating agent in that area: those plants that tend to flower at a certain time are more frequently pollinated and so this tendency will be passed on to the next generation. Thus a crucial interdependence between pollinator and plant is likely to develop over a comparatively short time, the plant requiring the pollinator for fertilisation, the pollinator needing the plant's nectar to feed on during a period when its usual food source is absent. Indeed the flowering of mangroves (especially if it coincides with a non-flowering period for neighbouring terrestrial plants) doubtless attracts numerous animals into the mangal that would not otherwise enter it.

Birds are thought to pollinate the flowers of the large-leafed varieties of knee-rooted mangroves. I have seen red-headed honeyeaters (*Myzomela erythrocephala*) feeding in the flowers of the large-leafed orange mangrove in the Northern Territory. The flowers of this mangrove are red which is thought to entice birds. This bird has also been observed feeding in the flowers of the closely related yellow-flowered knee-rooted mangrove (*Bruguiera exaristata*).

Other visitors to mangrove flowers (and presumed pollinators of them)

Butterflies are thought to pollinate the small-leafed varieties of knee-rooted mangrove. This butterfly from a Northern Territory mangal is Hypolimnas bolina nerina.

include butterflies, flies, bees and wasps; the latter two are the most important.

Not all mangroves are pollinated by animals however. Tomlinson has argued that the prop-rooted mangroves are probably wind pollinated. Wind pollination means that the plant in question does not need to compete with others for the available animal vectors in the habitat. There are virtually no insect surveys within mangals let alone surveys of seasonal fluctuation that could be correlated with flowering mangroves, but it is not difficult to imagine situations in which a mangal isolated from other vegetation may suffer from a dearth of visiting pollinators. Just as we saw earlier how seasonal flowering may be a response to the availability of a specific pollinator, perennial flowering may be a response to a general dearth of pollinating agents.

A plant that is wind-pollinated needs to produce a considerably larger amount of pollen than one that is not because the method is extremely imprecise and therefore wasteful. Very high pollen production is a feature of the prop-rooted mangroves. Furthemore, in these species the flower lasts only about 12 hours before the petals and stamens are shed. The flowers also tend to point downwards, making an awkward platform for animal visitors. Indeed the structure of the flowers of the prop-rooted mangroves are quite different from those of other members of the tribe such as the knee-rooted mangroves and the yellow mangrove. The difference in form ineluctably suggests a difference in function. Furthermore the prop-rooted mangroves do not produce either the nectar or the odour usually required to attract animal visitors.

Other plants of the mangal

In a habitat subject to such extremely complex fluctuations in conditions, one can occasionally come across plants that one would not expect to see in the mangal. For example I recently entered a mangal in Kimbe Bay, West New Britain, Papua New Guinea, confident in my ability to identify the majority of the mangroves I was going to encounter. This confidence derived from my familiarity with the mangroves of Australia's Northern Territory and I expected there to be a considerable overlap of species between the two locations. From the sea everything seemed to be as I expected: tall stilt-rooted mangroves fringed the seaward edge. Fighting my way into the dense, shaded interior I entered the domain of the knee-rooted mangroves. They towered far overhead and cast the ground into an almost perpetual shade. The mud was broken by the innumerable bumps of knee roots. I spotted a skink posing on a knee root. It was aglow with spectacular rainbow iridescences, but it darted away long before I was able to photograph it. As I advanced towards the landward edge I found a stream that was bringing fresh water down to the sea: several immense cannonball mangroves stood along its banks. This is a species of woody-fruited mangrove that is virtually absent from the arid regions of the Northern Territory because of its need for a regular supply of fresh water.

As I looked deeper into the knee-rooted zone dominated by large-leafed orange mangroves, my expectations received a rude shock. Exotic-looking vines meandered across the muddy ground over the exposed knee roots. Lush, huge-leaved epiphytes wound their way around branches and mangrove saplings. And there were ferns everywhere. Many were the well-known mangrove fern but other, clumping varieties were common and, stranger still, there were even some *Cyathea* tree ferns, not to mention several varieties of lush palm including *Pandanus* palms. My understanding of what constitutes a mangrove ecosystem was severely challenged.

Clearly this was an atypical habitat. There was evidence of regular felling of the large-leafed orange mangroves; there is a village close by and the wood is used in house construction. Where the trees had been cut the mangrove ferns were at their densest. Nevertheless the invasion by rainforest plants could only be explained by the fact that this area had, for some reason, had its salt water input blocked. The frequent rains in the region would help lower the salinity of the soil but that would not explain the presence of rainforest plants growing amongst the mangroves. Perhaps there had been a slight change in the height of the land relative to the sea level and much of the area was now above the tidal range. Or it could be that isolated mounds of sediment had been deposited above the tidal range and these had now been colonised by rainforest plants. The mud lobster *Thalassina anomala* builds such mounds

and, if high enough, these can become micro-habitats which support non-mangrove plants. Again, if the area was receiving a sufficient supply of sediment washed down from inland then the shoreline could be accreting. If the process was gradual enough then the true mangroves would presumably be able to adapt their root systems accordingly and last for many years. In other words the large-leafed orange mangroves might not now be in an intertidal setting. This species is known to be able to survive outside the mangal and perhaps it is capable of thriving in such a habitat once it has reached a certain stage of development. Indeed, the fact that large-leafed orange mangroves have been found growing and apparently thriving above the highest tidal limit raises several questions. The first is how the seedlings become established in an area which the sea – the medium that transports them – cannot reach. Presumably unusual weather conditions such as cyclones can cast the seedlings on to ground above the tidal level, but this raises another question, for this species at least: whether it really is at a competitive disadvantage with terrestrial plants in terrestrial habitats if it manages to establish itself there.

Opposite: This tree fern (Cyathea sp.), although growing among mangroves, is not a member of the mangrove community. It is presumably in an area where soil deposition has elevated the substrate above the level of the tide. True mangrove ferns (Acrostichum speciosum) are visible in the lower right of the picture.

The Animals
of the Mangrove Forest

The study of the animals found in the mangrove ecosystem is in its infancy. While numerous botanists have studied the plants to try to understand their adaptations for survival in the intertidal habitat, zoological studies have lagged far behind. Part of the reason is that many of the animals are not dependent on the mangal for survival. Although some spend their lives in the mangal, they do not appear to have to, as the same species also occur outside it; others may depend on the mangal at some stage of their development or use it for limited periods. Given the transient nature of much (though by no means all) of the mangal's animal population, zoologists interested in these transients are likely to choose less difficult environments in which to study them.

The old view that the mangal is a temporary habitat that will eventually be succeeded by a terrestrial one – usually a tropical forest – has only reinforced the prejudice that its animal population is likely therefore to be of secondary interest.

As biologists revise their understanding of the mangal and realise that it is better understood as an end in itself rather than a means to another end, so will both the plants and the animals become objects of considerably greater interest. Intuitively this approach is likely to be more rewarding. After all even if it were true that the mangal is only a precursor to the tropical forest biome, it is, when considered *en masse* a considerable community that has persisted in the tropical intertidal ecotone for many millions of years. A theory that considers mangroves at the local, short-term level cannot do justice to the intricacies of a biological system of considerable geological age.

The food webs that need to be constructed to understand the rôle of the mangrove fauna are likely to vary considerably in complexity depending on the species diversity of both plants and animals; an Indo-Pacific mangal that is composed of far more species of plant than its New World equivalent, would, one might expect, have a correspondingly greater variety of animals within it. This diversity could simply be a reflection of a wider range of animals available to invade the mangal or it could be due to the fact that the more extensive floristic assortment of eastern region mangals offers a larger variety of niche for animals. Again, the extent to which mangrove swamps are better understood as relatively long-lasting and stable communities will be reflected in the degree of specialisation of the animals in them. A modest example from my own exploration of a Northern Territory mangal suggests precisely such specialisation. On examination of the fluid-secreting developing flower buds of yellow mangroves I often

Assassin bug nymphs (Reduviidae) are fairly common on the developing flower buds of yellow mangroves in Darwin Harbour. The fact that their abdomens blended well with the buds suggests that they lie in wait to ambush visiting insects that come to drink the escaping sap.

found the nymphal stages of assassin bugs (Reduviidae) poised motionless on the buds. The coloration of the abdomen and the orientation of the body clearly suggested that the bugs were camouflaged and waiting inconspicuously for the approach of an insect that they could ambush, possibly one coming to feed on the escaping fluid from the flower buds of the plant.

There are several ways in which mangroves can offer animals normally associated with terrestrial forests an extension of habitat. A virtually seamless continuation from terrestrial forest canopy to mangrove canopy will allow animals to enter the mangal without 'knowing' that they are in the intertidal habitat. Furthermore the mangroves may provide animals with a corridor between two terrestrial forest habitats, or a refuge when their normal habitat is unsuitable – perhaps at those times when the food supply is temporarily inadequate there or the terrestrial habitat has been damaged by storm or fire. As I have said, where nectar-bearing mangroves flower during the non-flowering periods of their terrestrial plant neighbours, nectar-dependent animals are likely to invade the mangal in impressive numbers.

Obviously the above examples have little direct relevance to the fact that the temporary habitat is intertidal. When one considers animals specifically adapted to the intertidal zone one must distinguish between those that are found on intertidal mudflats and those that only occur on mudflats in or in the immediate vicinity of mangroves. The distinction is not always clear cut however as many mudflat

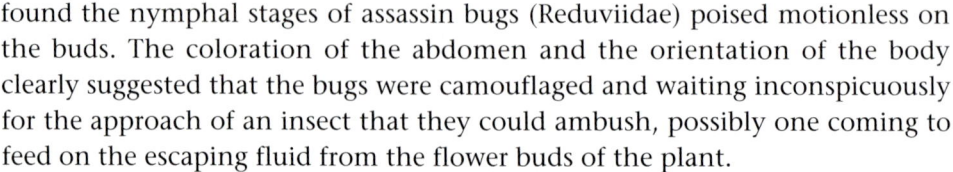

*Radjah shelducks (*Tadorna radjah) *feed on mangrove mudflats. These birds are commonly associated with tidal mudflats as well as paperbark swamps in Australia.*

A caterpillar (Lymnantriidae) feeds on a large-leafed orange mangrove leaf in Papua New Guinea. Several of these mangroves were heavily infested (and damaged) by these feeding caterpillars while other prop-rooted mangroves nearby were untouched. This might suggest that the prop-rooted mangroves have some chemical defence against these insects.

dwellers may be dependent on mangrove-derived leaf litter or detritus in which case it will only be a question of their being somewhere where the food source can reach them. For example about 70 species of fishes are recorded from Australian mangrove communities but only four species of mudskipper are thought to be specifically confined to this ecosystem.

There is a wide range of habitat available within the mangal, including the mangrove-fringed edges of rivers, the channels and creeks and the pools of trapped water formed in depressions when the tide has gone out.

Variations in the tree species will also be reflected in the fauna. For example different types of bark will be more vulnerable to boring beetles. The rich tannin content of the prop-rooted mangroves is presumably to deter beetles, and the bark-shedding capacity of the cedar mangrove may serve an equivalent protective function. Similarly, defensive chemicals in the leaves of different mangrove species will deter different herbivores. I remember several large-leafed orange mangrove trees in a Papuan mangal that were heavily damaged by feeding caterpillars while neighbouring stilt-rooted mangroves were untouched.

The flooding tide can bring marine creatures into the mangroves to feed,

The tree-climbing crab Aratis pisonii *is well capable of scuttling quickly up and down the prop roots of this red mangrove in Caroni Swamp, Trinidad.*

When the tide flows into the Bimini lagoon, numerous large southern stingrays (Dasyatis americana) *take up position in the mangrove trough zone: presumably to ambush any unwary crustaceans.*

shelter or spawn, while the receding tide will expose food-rich mudflats for animals from within the mangal or beyond it to feed on. Again, many of the animals that hide within the substrate when the tide is in are dependent on low tide to feed or reproduce.

Variations in the particulate composition of the substrate can determine the creatures found in it because of their ability to move through it and also because many of them have filtering processes for feeding that require substrate particles to be of a certain size. An example is the numerous species of fiddler crab which are only found in certain specific areas of the mangal because of the composition of the substrate.

Having made these general points it is time to look more closely at the animals of the mangal. There are two ways in which we can do this: either we can discuss the particular part of the habitat and then looking at the animals in it (for example the tree canopy, the trunk and root systems or the substrate); or we can look at each group of animals and discuss it in relation to the mangal. Since there is considerable overlap between the habitats and the animals, occasionally at a general and often at a specific level (certain mudskippers and crabs even climb trees, a bird can move from canopy to mudflat), both methods are somewhat artificial. I have therefore decided to deal with the animals by group, moving from one part of the ecosystem to another when appropriate.

Spiders and insects

The spiders and insects are among the least studied of mangal animals. There are some wonderful spiders in the mangroves. I remember explaining to a tourist in Tobago that I was working in the mangroves. He gave me a sorry look as if to say 'With all these beautiful beaches and sunshine and bikini-clad women you spend your time in the bug-infested swamp?' It was my chance to convince another person of the strange splendours of the mangal.

'I'll take you in and show you around if you like,' I suggested. He looked at me with a pitying expression. I thought I had better add something specific to convince him that it would be worthwhile. My mind returned to the huge spider with its 2 m wide web I had seen earlier in the day.

'You wouldn't believe the size of some of the spiders in there!' I added.

He shivered. 'Oh yes I would,' he said and fled!

The spider I was trying to interest him in is called the golden silk spider

(*Nephila clavipes*). It is a widespread species, ranging from the southern United States down into the tropics and is by no means confined to mangroves. Although I had not seen any in the Caroni Swamp in Trinidad, they were very common (as were several other species of spider) at the landward edge of the Kilgwyn Swamp of Tobago.

The female golden silk spider builds a huge orb web and positions herself in the centre. Given the size of the web, if the sun hits it at the right angle, it and its builder are conspicuous and easily avoided. However if you are not watching where you are going or the web is in shade you can blunder into it. The largest spiders also tend to string their vast webs across precisely the gaps in the trees where you would not expect to see them. After struggling through closely-packed trees you find a gap and naturally accelerate – straight into a huge web. The strength of the web is remarkable: it feels as if you have walked into an invisible barrier. When you see the size of the spider in the centre, you are likely to try to disentangle yourself from the web with an alacrity tinged with panic. The largest female golden silk spider I saw had a body length of about 6 cm; the total length, including the legs, was probably about 10 cm.

The male is considerably smaller than the female; perhaps about a centimetre in total length. He lives in the female's web and mates with her at his peril: she is more than likely to devour him before, during or after mating. The last option makes good sense for the female – she gains an extra food source to nourish her eggs, and the male probably dies soon after mating anyway. Nevertheless when I saw the male spider darting nervously around the periphery of the web, I could not help thinking that he would have benefited from a crash course in stress management.

The vast web of the golden silk spider is a micro-habitat in its own right. While examining the huge web I noticed several other, tiny spiders in it.

Above: A flesh fly (Sarcophagidae) has blundered into the huge web of a female golden silk spider (Nephila clavipes) in the Kilgwyn Swamp, Tobago.

Right: Its struggles send vibrations through the web and soon attract the spider, which immediately immobilises the fly by trussing it up in silk.

Opposite page, left: The female spider injects digestive fluids into the fly and then proceeds to feed by sucking out the dissolved body fluids. The considerably smaller male golden silk spider gingerly approaches to try to feed. The female rocks her body up and down and sends waves through the web that warn the male to keep away.

Below: Female kleptoparasitic spiders (Argyrodes elevatus) are less easily deterred than the male golden silk spider. They live in the web and specialise in stealing meals from captured insects. Although the female golden silk spider tries to knock them off the web with her legs they deftly sneak around the other side of the rapidly disintegrating fly to share in the meal.

These belong to the genus *Argyrodes*, and are termed kleptoparasites: they steal food from the web of the golden silk spider.

When I came back to the web a few minutes later a flesh fly (Sarcophagidae) had flown into it and was struggling to escape. The vibrations of its struggle soon brought the female golden silk spider down upon it and she wrapped the hapless fly in silk from her spinnerets. All this activity soon brought the male spider scurrying towards the food-source but the female was not going to share her meal with him unless she had to. She used one of her legs to try to flick the male away. She even rocked menacingly towards him and he retreated. But the tiny kleptoparasitic *Argyrodes* spiders were not so easily deterred and approached the captured fly from the other side of the web. The female silk spider tried to flick them away with her legs but they avoided her attempts. Then she started rapidly moving up and down in the web, setting up an impressive vibration that shook the *Argyrodes* spiders almost off the web. They retreated above the trussed fly while the female golden silk spider sank her huge mouthparts into it and started to fill it with the acidic secretions that would turn its innards to fluid.

While she was thus engaged the *Argyrodes* spiders produced their own thread lines and dangled their way down to the other side of the fly. The golden silk spider again attempted to flick them away a couple of times but then gave up and the kleptoparasites fed on the other side of the rapidly shrivelling flesh fly. After only a couple of minutes all that was left of it was a black, shrivelled mass of undissolved exoskeleton. By now the *Argyrodes* spiders had retreated. The female golden silk spider carefully removed the remains of the fly from her web and attaching it to a thread, lowered it out of the web. When it was free of the web she cut the line and the remains dropped to the ground. As soon as she had completed the task another fly blundered into the web and the entire process was repeated. It was easy to see how this particular spider had grown so huge.

The distribution of spiders within the mangal is often patchy. One might spend hours without seeing any and then come across an area (normally at the landward edge) dense with webs. In the Northern Territory many different groups of spiders can be encountered. One of the more conspicuous would be the unidentified funnel-building spider, which builds a funnel-shaped web leading into a hollow in dead wood. Here the spider waits to pounce on anything that becomes trapped in the funnel. There are also orb-web spiders (*Nephila* spp.) and jumping spiders (Salticidae) and doubtless numerous others.

The majority of spiders have probably wandered in from terrestrial habitats. A torch-lit examination of the mangrove mud at night might well reveal a scurrying wolf spider (Lycosidae). A wolf spider of the genus *Pardosa* has been recorded as feeding on juvenile fiddler crabs in a Malaysian mangrove swamp. It is thought to enter air-filled burrows before the tide returns. A brood of juveniles has also been found in an abandoned fiddler crab burrow.

*The gasteracanthid spiders (*Gasteracantha *spp.) can be very common in New World mangals. Here, in Kilgwyn Swamp, Tobago, a female gasteracanthid feeds on a large carpenter bee (Apidae). Subsequent trips to the swamp revealed that the meal lasted several days.*

The thick, hairy coat of the wolf spider is water-repellent, and this species is an example of a spider that is a full-blown mangrove resident.

The gasteracanthid spiders, with their characteristically spiny and laterally elongated abdomens, are common in both eastern and western hemisphere mangals. While exploring a mangal in Bimini, in the Bahamas, it was almost impossible to avoid meeting a gasteracanthid in its web every few metres. Their curious body shape may make it difficult for birds to feed on them. While I was photographing a gasteracanthid in the Bimini mangrove it moved away

*The green anole (*Anolis carolinensis*) is quite capable of changing colour to match its background. Here, in the Bahamas, one attempts to swallow the spiky mouthful that is a gasteracanthid spider. The lizard soon scurried off with the spider in its mouth so I could not see whether it succeeded in eating it.*

from its web towards a branch. An anolid lizard immediately rushed over and grabbed it in its mouth and then attempted to chew on it. It was rather like watching someone chew on a mouthful of glass. The lizard would move the spider into a different position and then try again. It seemed unlikely that it could actually swallow so spiky a mouthful without getting it lodged in its throat. I waited to see if the lizard would succeed in eating the spider but unfortunately the lizard chose to continue its chewing undisturbed and scuttled away up the tree with the spider still lodged in its mouth.

The most spectacularly coloured spider I have ever seen was in the mangroves of the Adelaide River in the Northern Territory. Russell Hanley and I went deep up a meandering creek in the Northern Territory Museum's little 'tinny' boat and then entered the mangroves at a relatively accessible spot. I found this spectacularly patterned and tiny (about 1 cm wide) gasteracanthid that had built her web deep within the mangroves – in fact in the stilt-

The spectacular Gasteracantha westringi *from deep in the mangroves of the Adelaide River, Northern Territory. The spikes may be designed to make the spider difficult to swallow; the bright colours may warn that it is a foul-tasting mouthful.*

rooted mangrove zone. There was a breeze, just enough to make the spider move in the web, and I could not photograph it because the depth of field was considerably less than the movement of the spider, owing to the wind. I was becoming frustrated when Russell called me over to show me something.

'I can't come just yet. I'm trying to photograph a spider,' I called.

'Oh don't worry about that. Wait 'til you see the one I've found here.'

'It can't be as spectacular as this one.'

'It is,' he replied.

'Is it sort of green, fringed with yellow, with orange spots and dark spikes everywhere?' I called.

'Yes, but more so. And you don't have to shout. I'm only 10 ft away.'

I struggled through the prop-root barriers of the mangroves and reached him, absolutely shattered, ten minutes later. Sure enough it was another of the beauties, and a close scrutiny revealed a few more in the immediate vicinity. We decided to collect one and take it back to the museum for identification. (We were subsequently informed that it was a *Gasteracantha westringi).*

Later that day, when we were back at our campsite, I set the spider up on some stilt-rooted mangrove leaves so that I could photograph it properly. Below us some fishermen were pottering along in their boat and we paid them no attention. The close-up outfit I was using did however happen to look like a telephoto lens. A few hours later one of the fishermen from the boat came over to our campsite and after some nervous small-talk about the weather (hot) and the crocodiles (everywhere) he came to the point.

'I was just wondering, mate, why you were photographing us in our boat.'

'Photographing you in your boat?' I asked.

'I mean it's not as if we're doing anything wrong fishing here is it?'

'What do you mean?' Russell asked.

'I mean I admit it,' he said. 'I am already due for one court appearance for fishing out of season. And my new fishing licence is in the post. And we haven't gone over the quotas or anything. So you won't show anyone those pictures will you – of us fishing I mean?'

'But he wasn't taking your picture, mate. He was photographing a spider. I'm a biologist and he's a photographer. We aren't interested in what you are doing. We are studying the mangroves,' Russell explained.

'So even if you took a picture of us fishing by mistake you wouldn't show it to anyone?'

'No, not a soul,' I promised.

He was all smiles and after a few more pleasantries about the weather (still hot) and the crocodiles (still everywhere) he left.

This incident highlights a message which bears remembering. Many mangrove forests are in remote, seldom visited locations. They are vulnerable to over-exploitation and fishing regulations mean little to certain individuals (although not, I am sure, this particular fisherman) if there is no one to enforce the regulations.

The insect fauna of mangroves is incredibly diverse but it is virtually unstudied. The vast majority of the arboreal species presumably wander in from neighbouring terrestrial forests. Amongst the more conspicuous insects likely to be spotted are bees and wasps (especially when the mangroves are in flower), dragonflies, mayflies, termites, grasshoppers, flies, beetles, bugs, butterflies, moths and ants.

Two types of ant merit further discussion. The leaf-cutter ants of central America (*Atta* spp.) cut leaves and flowers from plants and take them down into underground nests. Here the plant material is ground into pulp. It is then mixed with the ants' droppings. Next, the vegetative parts of a specific fungus

cultivated in the nest are introduced. The fungus flourishes and the ant colony lives entirely on this food-source.

While inspecting a black mangrove tree in the Caroni Swamp I spotted a procession of leaf-cutter ants each carrying a fragment of leaf or flower down the trunk of the tree. I decided to follow them to see where the nest was: clearly, since the nest is underground, it could not be within the intertidal zone. Sure enough there was a place where the prop root of a red mangrove touched the black mangrove tree. At this point the ant procession left the black mangrove and marched in a seemingly endless line along the prop root of the red mangrove, across its trunk and along another prop root. The direction was landward. Near the end of this tree's most landward prop root was another red mangrove tree and there ants continued along the red mangrove prop roots for some 20 m until they reached the bank of higher ground. Then they marched along the ground about

These vespid wasps (Ropalidia gregaria) *built their nest on the branch of a grey mangrove tree in the Northern Territory. Wasp nests are by no means an uncommon sight in the mangal: the majority of mangroves require insects for pollination and produce nectar to attract them; wasps are among the most important pollinating agents.*

5 m before disappearing into a hole in the ground which led to the nest. This was a noteworthy example of incursion into the mangroves by terrestrial animals that know precisely how to avoid the inconvenience of the tides!

Another interesting variety of ant, the weaver ant (*Oecophylla smaragdina*), is found in the Northern Territory mangal. Workers bond the leaves of a plant together by gripping one and holding another in the mouthparts. Leaves can thus be temporarily stitched together. The larvae are then carried by other workers to the edges of the leaves where they are stimulated to secrete a silk-like material. The workers move the larvae back and forth and so bond the leaves together. In some species of mangrove, such as the yellow mangrove, several leaves are glued together to form the new nest. In the native hibiscus, a tree common at the landward edge of Northern Territory mangals, the ants will simply fold one leaf

*This leaf-cutter ant (*Atta *sp.) carries a black mangrove leaf along a red mangrove's prop root. Note the termite gallery trail on the underside of the prop root (termites avoid sunlight). Leaf-cutter ants can enter the mangroves by gingerly keeping to routes above the tidal height. They cut leaves and flowers from trees and then store them in their subterranean nest where they are used to cultivate the fungus on which the ants feed.*

Below: Australian weaver ants (Oecophylla smaragdina) build their nest in several stages. First, several ants clamp their jaws into one leaf while gripping another.

Right: Then the larvae are carried to the joint and stimulated to secrete a silk-like material that glues the two surfaces together.

Left: This is a completed nest made out of several yellow mangrove leaves.

Above: In the case of the native hibiscus (Hibiscus tiliaceus) a mangrove associate, a single leaf is simply folded back on itself; the entry point is where the apex of the leaf meets the stem.

Dragonflies are a common feature of the mangrove insect fauna. Many have nymphal stages that can tolerate brackish conditions. Here a common skimmer (Libellulidae) rests on a twig in a Bahamian mangrove.

back onto itself and glue the edges together: the width of the leaf and its shape allows it to fit perfectly.

Certain dragonflies and mayflies (Odonata) as well as caddisflies (Trichoptera) have nymphs which develop in pools of brackish water. But the list of insects which can tolerate sea water for at least a part of their life cycle is very short. Certain mosquitoes (Culicidae) and biting midges (Ceratopogonidae) are all too painfully obvious. The latter lay their eggs in the mud, especially in the openings of crab burrows. There are also midges (Chironomidae) which feed on mangrove detritus. Other insects of the intertidal realm include the unloved, biting horseflies (Tabanidae), as well as a few other groups of thankfully less noticeable flies.

The water-striders (Gerridae) are the only group of insects that have marine representatives. Of the 44 known species the majority are coastal in their distribution while five are found on the surface of the open ocean. The coastal species require warm, sheltered conditions and are common in mangrove swamps. At high tide they can be seen darting along the surface between the trees and the root systems. At low tide they occasionally congregate in high numbers in pools of sea water. They are carnivorous and feed on other insects which fall into the sea and also devour small marine animals – zooplankton and larvae which they grab with their powerful front legs. The victim is injected with digestive fluids via the tubular mouthpart. The water-strider then sucks out the resultant soup.

The reason for the relative paucity of marine insects is not well understood.

The water-striders (Halobates spp.*) – photographed here in an Australian mangal – are the only insects capable of spending their entire life cycle in sea water. They skim along the surface in mangroves hunting for insects that fall into the water as well as devouring planktonic animals in the water column below them.*

Mangrove molluscs

Molluscs are a conspicuous constituent of the mangrove fauna. Even a cursory investigation of the branches of the trees will show them to be dense with periwinkle snails; some are prettily patterned, others more dull in colour. The patterning usually reflects the background on which they are found: a dark snail on a dark background, a light snail on a light background. Camouflage reflects the fact that they are preyed on by birds.

It used to be thought that one species of periwinkle snail (*Littorina scabra*) was dominant in the trees of Indo-Pacific mangals over vast distances. Now however the supposed species has been re-examined, renamed and divided into 20 different species of the genus *Littoraria*. The most common species in Northern Territory mangroves is *L. filosa*.

Mangrove periwinkles are found above the influence of the tide. The sexes are separate and mating occurs in moist conditions; for instance after rain, just after high tide, or early in the morning. The males crawl rapidly along in search of another snail of the same species and mount any they encounter. If it is a female then copulation takes place; if it is a male they beat a hasty retreat. Marine and aquatic molluscs normally produce a veliger larval stage which develops in water. However many of the species of mangrove periwinkle stay above the water level and so have had to alter their reproductive strategy. One solution is for the female to carry the fertilised eggs in a special pouch; they develop within her into tiny snails.

Mangrove periwinkles feed on the surface film of microscopic algae growing on the leaves, branches and trunks of the mangrove trees. It has been suggested that they may be in the process of evolving from the marine to the terrestrial condition.

A very important group of molluscs in the mangal are the bivalve molluscs normally referred to as shipworms (*Teredo* spp.), which are better known for the damage they cause to the hulls of wooden ships. Shipworms use their considerably reduced shells as tools for burrowing into wood. As they progress they secrete a layer of lime around themselves, thus forming the familiar limestone tubes so often seen in the dead wood cast ashore after storms. The animal feeds partly by ingesting some of the wood it breaks down but also by ingesting food items taken in from the sea water; it discharges waste products

Top left: The periwinkle snails of Indo-Pacific mangrove swamps have recently been separated into numerous new species. These, from a Northern Territory mangal are probably Littoraria filosa. *Top right: The periwinkle snail on the left is a male. It is rapidly (for a snail) approaching the snail on the right. Bottom left: On catching up with it, the male attempts to copulate. Bottom right: On this occasion the male continued to crawl all over the other one suggesting that it was a female; if it had been a male the first snail would have made off.*

back out through the tunnel. The significance of shipworms in the mangal is as the first step in breaking down dead wood into smaller units that can be attacked by other organisms. One might think that they have something of a charmed existence in the mangal given that there is an almost limitless supply of wood. However a close examination of a *Teredo*-damaged log in an Australian mangal is likely to reveal several murex shells on it. The mangrove murex (*Naquetia capucina*) feeds on shipworms, and uses its extremely long proboscis to probe for them in their limestone tunnels. The mangrove murex also deposits its eggs in the tunnels of the dead worms.

At low tide, an inspection of the muddy floor of a Northern Territory mangal may reveal numerous flattened, inconspicuous, slug-like animals. These are little-studied pulmonate (air-breathing) molluscs belonging to the genus *Onchidium*. They feed on detritus on the mud surface and must crawl up trees to avoid the incoming tide. This ability to anticipate the tidal cycle is also found in a small creeper snail (*Cerithidium anticipata*) which feeds

*This photograph taken on a red mangrove tree in the Bahamas shows how mangrove periwinkles (*Littorina angulifera*) avoid submergence by climbing up the prop roots as the tide comes in.*

Shipworms (Teredo *spp.*) are very important in the mangroves because their boring activity initiates the breakdown of dead wood.

The mangrove murex (Nequetia capucina) *has a very long proboscis and predates shipworms by extending it far down their tubes. It also lays its eggs in abandoned shipworm tubes.*

The surface of the mud in Northern Territory mangals can have numerous small slug-like animals (up to about 5 cm) slithering across it. These air-breathing molluscs belong to the genus Onchidium. *There are probably several species, although as yet they have not been much studied. They feed by ingesting the surface layer of mud and removing detritus, bacteria and algae from it.*

on tiny food particles on the mud when the tide is out and then climbs up a tree to safety before being inundated. While most animals have mechanisms which enable them to thrive either in the daytime or at night, animals living within the intertidal zone need another internal clock to synchronise their activity with the tidal cycle. Something as slow moving as a snail that has to climb a tree to avoid the tide has to set out on its return journey long before the tide reaches it or it will be swamped. Thus not only will an intertidal animal be likely to have an internal clock regulating its activity with the sun, it will also have another clock for synchronising its activity with the lunar cycle, which is the primary cause of tidal fluctuations.

In most parts of the world two tidal cycles will occur in a solar day. The sun has less influence on the tide than the moon. However when the moon is new, and also when it is full, the sun reinforces the gravitational pull on the Earth causing spring tides twice a month. Thus within the intertidal domain are creatures whose internal clocks synchronise their behaviour to the frequency and duration of tidal influences in that specific position in the intertidal zone. More remarkable still, the solar and the lunar clocks must work together so that behaviours such as feeding and mating are preset to occur not only during the right part of the tidal cycle but also during the appropriate part of the day. These internal clock mechanisms are also found in intertidal plants. Certain microscopic plants on the mud's surface are known to bury themselves before the rising tide reaches them. Presumably the respirational activity of the mangrove plants (including the opening and closing of the lenticels) is similarly co-ordinated.

*These creeper snails (*Cerithidium anticipata*) appear to be able to anticipate the incoming tide; they crawl up trees before it reaches them. The purpose of this is thought to be to avoid predation when the tide is in; they may also leave the mud to avoid desiccation when the mudflats dry out.*

The intertidal mudflats of a mangal can be subjected to extremes of temperature. For example a wet, shaded area will be considerably cooler than a sun-baked dry one. At the rear of a Northern Territory mangal it is common to find large potamidid mud whelks, up to about 10 cm in length, crawling around on the surface mud at low tide. Some will be out in the open while others will be clustered under low mangrove trees apparently seeking their shade from the worst of the heat. Closer examination will probably reveal that there are several species involved: *Telescopium telescopium* (with a fairly smooth though finely grooved shell) as well as two similar-looking species of mud whelk (*Terebralia palustris* and *Terebralia semistriata*) whose shells are more scalloped in outline.

Robert McMahon, a biologist from the University of Texas at Arlington, is interested in thermal tolerance in molluscs and has made a study of these mangrove whelks to learn more about how they manage to survive the often brutal temperature range founded within their habitat. He noted that *Telescopium*

Left: The mud whelk Terebralia palustris feeds on fallen leaves on the mangal floor. Right: It is common to see mud whelks form aggregations in the shade of trees when the tide has gone out. Experiments show that they can tolerate amazingly high temperatures on the mangrove mudflats. Despite appearances to the contrary such aggregations may simply reflect where dead leaves are most likely to fall.

telescopium can be found wandering around almost anywhere in both shade and sun. It appears that the creature feeds by ingesting the surface material and digesting the detritus in it. If the substrate on which it is feeding becomes dry, however, the animal seals itself into its shell and awaits the return of the tide. *Terebralia palustris* forms aggregations, usually in the shade under the trees. It appears to feed on recently fallen mangrove leaves. Meanwhile *Terebralia semistriata* is something of a wanderer, although usually in areas at least partly shaded by the tree canopy. It appears to be a generalised detritivore in the same mould as *Telescopium telescopium*; it also retreats into its shell if the substrate dries out. One wonders whether the feeding methods of the whelks are determined by their ability to withstand temperature, and whether less temperature-tolerant species need to seek shade to reduce water loss due to evaporation when the tide is out during the hottest part of the day.

Measurements of the body temperature of *Telescopium* snails by Professor

McMahon demonstrated that they can be up to 3°C higher than the substrate temperature, which ranged from 32 to 34°C. However, the shaded *Terebralia palustris* snails have body temperatures averaging 2 to 3°C less than the sun-exposed *Telescopium*. This would suggest that shade keeps the animals cooler. However Professor McMahon tested the thermal tolerance of the snails by artificially exposing them to higher temperatures and discovered that all three species could survive at up to 46°C and, astonishingly, all three were even able to survive limited exposure to temperatures in excess of 52°C. This remarkably high degree of temperature tolerance is an impressive adaptation to the extreme heat that they must regularly endure in the tropical intertidal habitat. Presumably *Terebralia palustris* is found under the trees (and fortuitously in the shade) because that is where fallen leaves are most likely to be found.

There are other kinds of mollusc to be found in the Northern Territory mangal. The mud snail *Nassarius dorsatus* is common in the lower mangrove zone and will appear as if from nowhere (in fact under the substrate) to scavenge on dead animals.

Various species of oyster (*Crassostrea* spp.) are commonly found attached in considerable numbers to the roots of prop-rooted mangroves. The tide will often scoop out a trough of deeper water immediately in front of a bank of prop-rooted mangroves, which will be invaded by some of the mangrove roots and will rarely if ever be exposed by the tide. Such roots can become festooned with oysters which then contribute to the establishment of a micro-habitat for numerous other creatures, including mussels, barnacles and worms. Mangrove oysters are harvested in many parts of the world, but a moment's thoughtless harvesting can cause considerable damage. For example, the roots of the red mangroves in Trinidad's Caroni Swamp hold vast supplies of mangrove oysters (*C. rhizophorae*). To pick each adult oyster from a root is difficult and time-

consuming work. All too often a fisherman will simply cut the root off the tree and haul all of the creatures attached to it into his boat. This kills the immature oysters and everything else on the root, and is also very damaging to the tree. If too many roots are cut thus the whole tree could die.

The crustaceans of the mangrove

The crustacean fauna of the mangal is amongst the most conspicuous. While various types of crab are the most obvious on exposed mudflats they are by no means the only crustaceans in this ecosystem.

I will begin by considering an important, if rarely observed inhabitant of the crustacean mangrove community. An exploration of the habitat anywhere from India across to Fiji and Samoa may well reveal numerous conspicuous mounds of mud ranging in height from perhaps 20 cm to 1 m or more. These are most conspicuous at the upper reaches of the intertidal zone where the tide erodes them less

*The mud lobster (*Thalassina anomala*) is a common though rarely seen inhabitant of northern Australian mangroves. It spends most of its time under the mud in its extensive burrow. The burrow-building activity brings deep mud to the surface and can make otherwise buried nutrients available to the mangrove plants.*

frequently. They are built by the secretive and little studied mud lobster *Thalassina anomala* which extends above the intertidal level as well.

The mud lobster typically grows to about 20 cm in length (30 cm seems to be the maximum size). Its burrow is estimated to reach a depth of some 2 m. Because it rarely leaves its burrow, little is known about its natural history.

Where mud lobsters are plentiful they can have several important effects on the mangrove habitat. They bring up mud from beneath the level of the mangrove roots and deposit it in mounds on the surface. This activity makes available to the mangrove plants nutrients which would otherwise be too deep for them to reach. Furthermore the mound-building activity actually creates micro-habitats which are above the height of the surrounding area. Other species of plant, such as the mangrove fern, can become associated with these artificially elevated areas.

Mud lobsters are thought to be mud eaters: they ingest mud and filter out organic material from it. They are scrawny, with little meat on them and are of little commercial value. However their burrowing activities can cause them to be considered a pest. Prawn ponds built in areas of cleared mangroves can be damaged by their burrowing activities.

Mud lobsters constantly shed their exoskeleton as they grow and leave it in the burrow. Fossilised mud lobster exoskeletons are quite commonly found in various places and are often incorrectly called fossilised mud scorpions.

The only time I have seen mud lobsters was when I was in a Northern Territory mangal walking along the immediate edge of the receding tide. I walked into an area where four small mud lobsters, about 8 cm long, suddenly emerged from the substrate, crawled around on top for a few minutes and then dug themselves back in at different entry points. I have no idea why!

Considerably more conspicuous and therefore more easily observed than

Male fiddler crabs (this is Uca flammula from the Northern Territory) use the smaller claw to pick particles from the mud, which are then passed into the buccal cavity. The larger claw is used for displays to attract females and deter rival males, for fighting with other males and as a sacrificial appendage that can be lost to a predator while the crab escapes.

mud lobsters are the fiddler crabs. They are by no means confined to mangrove areas; they also occur in temperate salt marshes as well as being able to survive beyond the intertidal range, for example on river banks.

Some fiddler crabs are drably coloured while others are so strikingly bright that they add dramatic spots of colour to an otherwise dull setting. One of the most vividly coloured is *Uca flammula*. This flame-orange specimen is among the most common Northern Territory fiddler crabs and is typically found on the soft mud of creek banks which are lined with stilt-rooted mangroves.

This species of fiddler crab (Uca capricornis) is common in areas of soft mud in Northern Territory mangals. The bright colours may be a species recognition signal: where more than one species occupies the same habitat it is important to recognise members of one's own species. Although the bright colours make fiddler crabs very conspicuous to bird predators, they have high-mounted eyes and presumably good all-round vision. If a bird swoops down on a group of fiddler crabs, the first one to spot it and dash for its burrow will initiate a rapid collective reaction.

The name fiddler crab derives from the fact that in the male crab one claw is grotesquely enlarged. There are thought to be over 60 species worldwide. Identification in the field can be difficult because of the colour range and variations in anatomy within species.

Examination of the stomach contents of fiddler crabs shows that they usually feed on both vascular plant tissue and algae, although animal matter has also occasionally been recorded. They feed on the surface of the intertidal mud by delicately plucking particles from it and ingesting them. In females both claws (chelae) are used. In males the gigantic claw is not used for feeding. One of the smaller species of fiddler crab has been observed to ingest grains of sand individually while the larger species will ingest correspondingly larger clumps. The ingested material is taken into the buccal cavity where it is cleaned of its organic matter. This is accomplished by bristle-like structures in the mouth parts. The matter is worked rather like a car being passed through the brushes of an automatic car wash: the car can be thought of as being the inorganic matter and the dirt removed from it as being the detritus.

Uca polita requires a coarse, sandy substrate, and is often found in front of Northern Territory mangroves.

One is often struck by the patchy distribution of the various species of fiddler crab in a mangrove swamp. A particular species may be very common in one area of soft mud, while a little further on where the substrate is coarser, a different species will be encountered. This distribution is a reflection of the adaptation of the mouth parts to feeding on different substrates. A fiddler crab which feeds on a coarse sand substrate will need mouth parts capable of removing organic material from the sides of sand grains. However, in a silty area a fiddler crab will take in a more soupy mixture of organic and inorganic matter. Separation of the two will involve other, finer, filtering mechanisms and also perhaps capillary action between mouth parts to hold the detrital material apart from the inorganic matter. It also appears that species adapted to more silty areas can ingest small amounts of inorganic material with their food.

One of the reasons why fiddler crabs cannot penetrate far into the terrestrial habitat is because the method of separating organic and inorganic matter requires water that is derived from the gill cavities. They need a constant supply of water to feed and also to keep their gill cavities saturated.

Fiddler crabs build burrows in which to retreat at high tide to avoid predation. It has been suggested that those that live in relatively dry parts of the intertidal range may be more territorial because their burrows provide them with a steady supply of vital water. If a fiddler crab is displaced from its burrow it risks dehydration. In certain species it is not unusual to see a wandering individual moving across the mudflat being aggressively challenged by other crabs guarding their burrows. It will challenge them and then, on finding a submissive one, displace it from its burrow, force its way in and disappear. One might think it is taking over the burrow but it reappears a few seconds later and continues on its way. It is presumably simply topping up its water supply.

Different species of fiddler crab can face very different problems of feeding. For instance a crab at the higher, open, rarely inundated areas of the mangal will presumably find the substrate relatively poor in organic matter, while one in a frequently inundated area rich in trees will find a steady supply of food. This raises the question of how hard they have to feed in order to survive. At the back of Ludmilla Creek, a mangrove area just outside Darwin, are open, sandy areas that are only covered by the higher tides. A very beautiful species

This beautiful fiddler crab (Uca elegans) is found on a firm, sandy substrate at the back of the mangal in areas not always inundated by the tide. It presumably spends long periods in a comatose state in its burrow to conserve energy and its emergences are pre-synchronised by internal clocks to the tidal cycle.

of fiddler crab (*U. elegans*) lives in this area. On many occasions I have walked across these flats when the tide has just gone out without seeing any fiddler crabs, although I have seen numerous burrows. On one occasion I found the flats swarming with these lovely little crabs, but on returning the next day when the tide had just gone out again only one was visible on the whole sand flat. Time and again while exploring Northern Territory mangals I would be surprised to see when the local fiddler crabs would emerge. One moment a mudflat would appear barren; a minute later hundreds of busy little fiddler crabs would appear at the same time. One day a mudflat would be bouncing with activity; the next, under apparently the same conditions, it would be a desert.

Fiddler crabs do not only emerge from their burrows at low tide in order to feed. They also indulge in elaborate signalling to each other, one aspect of which is the males waving their enlarged claw, to establish territories and also to persuade females to mate. It used to be thought that the large claw of the male is merely for display and is not used in combat. However I have occasionally seen various males indulge in brutal battles which include some impressive

Male Uca elegans *fighting. Fights between male fiddler crabs can be violent and result in one crab losing its major claw. Given the complex behavioural displays of the different species it would be interesting to know whether there is a correlation between the type, intensity and frequency of display and the tendency of a species actually to fight.*

judo-like throws and can result in the major claw being broken off.

Photographing fiddler crabs requires the ability to stay absolutely motionless. Time and again I have heard the crunching noise of battling fiddlers going on nearby, but one has to turn one's head to look in *very* slow motion to avoid startling the animals back into their burrows. Trying to bring the camera round to bear on a fight usually results in being in position after the fight has stopped; move any quicker and every crab will dart for its burrow.

The most visually impressive display I have seen by any fiddler crab is that

A still photograph cannot do justice to the elaborate display of this male Uca maracoani *of Caroni Swamp, Trinidad. The males raise and lower themselves on their legs while the large claw is moved in huge circular arcs. This ritual may minimise the inevitable injuries caused by fighting.*

performed by the male scissor fiddler crab (*Uca maracoani*) in Caroni Swamp, Trinidad. The greatly enlarged claw is opened and moved out and around to the side in large arcs; simultaneously the crab raises and lowers itself on its legs. The sight of hundreds of these crabs doing this on mudflats as far as the eye can see is spectacular.

Some species of fiddler crab have the ability to intensify their coloration. Furthermore studies by the scientist H. O. von Hagen in Ludmilla Creek in Darwin Harbour have shown that some females of the species *Uca polita* perform a waving display as well. Some 2.5 per cent of the females were observed to wave. They also had heightened colorations, and the combination presumably signals extreme desire to mate. Professor von Hagen noted that mature male and female *U. polita* often have burrows less than 20 cm from each other; he termed these pairs 'resident breeding units'. The male would wave-display at other males to keep them away while he approached his female neighbour and attempted to mate. She would retreat to her burrow and he would try to coax her out 'by interspersed sounds produced by his major cheliped' (the large claw). Copulation is induced by leg-stroking.

While a resident male will defend his female from other males, he will mate with any other female that he can. Thus he will court neighbouring females and also wander in search of others with whom to mate. Another strategy of certain males is to wait by the burrow for an occasional wandering female to come by and then to attempt to mate. The male will lure her into the mouth

Top left: The mating behaviour of a resident breeding pair of Northern Territory fiddler crabs (Uca hirsutimanus) is initiated by the male waving his large claw up and down. After several minutes the nervous female approaches.
Top right: After more bouts of male claw-waving the crabs go back to back and approach the female's burrow.
Centre: The female goes a little way into the burrow with the male following.
Bottom left: Mating occurs in the mouth of the burrow. The male waves his large claw several times, perhaps to deter rivals from approaching.
Bottom right:The male returns to his burrow (exhausted), with the female in pursuit expecting an encore.

Mating fiddler crabs (Uca flammula) in a Northern Territory mangal. Where different species of fiddler crab share the same area they appear to co-ordinate their mating timetables so that they do not overlap.

of his burrow and then strike a rather dramatic pose by raising himself high and waving his large claw as if signalling the impending conquest. He will aggressively approach his neighbouring breeding-unit female and make her retreat down her burrow so that he is not disturbed in his activities. On one occasion von Hagen noted that such a mistreated female was so upset that she went down her burrow, plugged it with mud and did not reappear for the rest of the period of low tide.

Clearly a fiddler crab which wishes to be brightly-hued to attract a mate must synchronise its activities according to daylight. The tidal cycle is usually 50 minutes later each day. This means that twice a month – every fifteen days – the tidal pattern will be the same. Experiments in which fiddler crabs were monitored in artificially induced tidal tanks have shown that they synchronise their displays and colour-intensifying behaviour so that they reach a height twice a month: on those periods when the height of the tide and the time of

A fiddler crab from a mudflat in a Northern Territory mangal. Some species show considerable colour changes, which are related to their breeding cycle and how recently the crab has moulted. This may be a colour variant of Uca capricornis.

day are at an optimum for the mating strategy of the species. More remarkable still, it has been shown that where more than one species occupies the same area they time their reproductive timetables so as to be out of synchronisation with each other.

There are areas higher up the intertidal zone which are not covered by the highest neap tides: in other words they will dry out for a few days at regular intervals. Fiddler crabs at this intertidal height are also able to adapt to this: they do not emerge to feed during this period. In fact when some were dug out of their burrows during the period, they were comatose, suggesting an energy-saving process at

work. A similar process is doubtless involved in fiddler crabs in areas which are only flooded during the periods of the spring tides, such as the beautiful *Uca elegans* mentioned earlier.

Although it is not understood how the crabs synchronise their activity to the intertidal domain, it has been shown that the mechanism is flexible. Fiddler crabs moved to areas with other tidal cycles will eventually adapt to them. Furthermore some intertidal species have populations living above the tidal range (for instance on the banks of rivers), which do not employ a lunar-cycle. Colour change as a precursor to mating has also been shown to vary

*Left: The bockadam (*Cerberus rhynchops*) of tropical Australia is a snake that has adapted to the mangroves. It tends to hunt at night and feeds on fiddler crabs and other crustaceans as well as on mudskippers. It will enter the burrow of a crab to extract the resident. The bockadam tends to cease activity as the day progresses and will bury itself in the mud to keep cool, with just its nostrils and eyes exposed.*

*Opposite: The mangrove or collared kingfisher (*Halcyon chloris*) is very common in some Northern Territory mangals. The large, upturned bill may be designed so that it can prise prey out of their burrows. It is sunset; this mangrove kingfisher has just caught a fiddler crab in the last few minutes of good light.*

within a single species along the gradient of the intertidal domain of a single beach: different heights on the same beach will have different durations of exposure and submergence and the differing colour-change timings reflect this.

Fiddler crabs build burrows which range in size, according to species, from a few centimetres to about 30 cm in length. The burrow is often curved in shape: after entering the mud vertically it bends until its long passage is parallel to the surface. They use their burrows to hide in at the approach of predators such as snakes and birds when the tide is out. As the tide returns a fiddler crab will re-enter its burrow and seal it with a plug of mud. Again, this is to hide from feeding fishes which enter the mangal on a rising tide. A wide range of animals is known to predate fiddler crabs. In a Malaysian mangal, mudskippers (*Periophthalmodon* and *Periophthalmus*), bockadam snakes (*Cerberus rhynchops*) and a species of wolf spider (*Pardosa* sp.) were recorded as regularly feeding on fiddler crabs. Seasonally present predators included a wide variety of birds such as the white-collared kingfisher (*Halcyon chloris humii*), the common sandpiper (*Actitis hypoleucos*), the redshank (*Tringa totanus*), the eastern whimbrel (*Numenius phaeopus variegatus*) and the long-tailed macaque (*Macaca fascicularis*). The range of birds predating fiddler crabs in other mangals is equally varied.

The greatly enlarged claw of mature male fiddler crabs offers some defence against predation either as a weapon, or, when brightly coloured and waved about, as a sacrificial diversion that can be lost to a predator and allow the crab the opportunity to escape down its burrow. Regeneration of the limb is possible.

Fiddler crabs are by no means the only crabs in the mangroves, nor are they the only burrow-builders. They belong to the ghost crab family (Ocypodidae) and at least 17 other genera of the family are known to be associated with mangroves. Also abundantly present are grapsid crabs (Grapsidae), although most are members of the sesarmid sub-family (Sesarminae). At least nine genera are recorded from mangroves. Various species of sesarmid crab (*Sesarma* spp.) are very common although inconspicuous. They tend to be omnivorous and will feed, among other things, on leaves that have fallen from the canopy, which they store in burrows in the mud. They leave their burrows and climb into the vegetation to avoid the incoming tide.

We have discussed how the burrow-building activities of the mud lobster is important because it brings nutrients up from the depths and also helps to aerate the mud. The same situation occurs with the burrows of burrowing crabs.

This grapsid crab (Metapograpsis frontalis) is an omnivore: it will eat leaf litter and also other crustaceans. It has dispatched a mantis shrimp.

*A male fiddler crab (*Uca dampieri*) emerges from its burrow in a Northern Territory mangal. The burrowing activity of crabs in mangrove forests is thought to increase the productivity of the trees as it aerates the soil and therefore reduces its toxicity.*

Experiments performed in Queensland mangroves have shown that aeration of the soil by crab burrows decreases its sulphide and ammonium toxicity. Furthermore the trees in the immediate vicinity of extensive crab burrows tend, when everything else is equal, to be the most productive.

Certain burrowing crabs might be considered to be permanent residents of the mangal. Those sesarmid crabs that are heavily dependent on fallen mangrove leaves for food cannot survive far from it. One could argue that if there is an area within the mangal which is populated by a variety of animals usually only found within it and their activities there significantly improve the productivity of the area, then they are both invaluable and dependent members of the mangrove fauna. Just such a situation has been shown to exist among the burrowing crabs in the stilt-root mangrove areas such as tidal creek banks in northern Queensland mangals. Not only does the burrowing activity increase forest productivity, but the sesarmid crabs are a crucial early stage in carbon recycling as they initiate the breakdown of mangrove leaves.

There are other groups of crustacean, such as the mud-living crustaceans like mantis shrimps, amphipods and pistol shrimps, and the shrimps that enter the mangal on the rising tide. The most commercially significant of these are the members of the genus *Penaeus* whose post-larval development requires the mangroves for food and shelter.

The mud crab *Scylla serrata* grows to a massive size and is caught commercially in mangrove areas throughout its range.

Mantis shrimps are aggressive and powerful burrow-building predators of shallow tropical waters. There are hundreds of different species and some are adapted to the muddy conditions found in mangroves. Their front pair of claws are immensely powerful and lightning-fast and are used to pierce or smash prey. This is Chloridopsis scorpio *from a Darwin Harbour mangrove.*

*The mudflats in and around mangroves are literally teeming with numerous varieties of virtually unstudied animal. These tiny opistobranch molluscs (*Haminoea *sp.) in a Darwin Harbour mangal are mating on the mud at low tide; they are probably hermaphrodite.*

Other invertebrates of the mud

The variety of invertebrates inhabiting the uppermost layer of mangal mud is very large, and includes polychaetes, nemertines, nematodes and sipunculids. Biologists in the Northern Territory have done surveys of the animals of the mangrove mud. It appears that the seaward edge of the mangal (the mangrove apple zone) is a transitional zone between the mangal habitat and the mudflats in front of it: juveniles from both habitats are competitively displaced by adults and so develop here. About 100 – 120 different species of mud-dwelling animal were found in the mangrove apple zone. Sampling further down the intertidal zone, out of the mangroves, showed that the number of animals increased because of the higher number of species that requires regular inundation. Sampling further up into the tidal creek bank mangrove areas, which are lined with stilt-rooted mangroves, showed that the number of animals remained fairly constant at about 100 species. Significantly the overlap with the mangrove apple zone was only about 15 per cent of the species while no appreciable overlap was found between the two outer zones. This reinforces the point about the unique range of animals found within the stilt-rooted mangrove zone.

The increased burrowing activity of animals in the tidal creek bank zone, together with (and stimulated by?) the higher tidal energy of the area tends to increase the productivity of the plants there. The higher productivity also leads to greater animal speciation. Similarly, faunal speciation in mangroves doubtless decreases with increasing aridity. Russell Hanley found that the number of animal species in Darwin Harbour mangroves, with an annual rainfall of about 1600 mm, is significantly greater than in similar mangrove communities in the more arid Borroloola area of the Gulf of Carpentaria, where the annual rainfall is about 600 mm.

Animals of the seaward trough zone

The activities of tide and current on the seaward edge of mangroves often scoops out a trough of deeper water immediately in front of and parallel to the trees. Anyone attempting to wade into the mangroves from the sea should beware – it can suddenly become a lot deeper just when one thinks one is there, and this is also an obvious hiding place for a crocodile.

This area is rarely if ever drained, and can be filled with the roots of various mangroves. Most typically it is prop-rooted mangrove roots that invade this

The activities of tide and current can gouge out a trough immediately in front of mangroves. Here, in Bimini, the Bahamas, the prop roots of red mangroves which have invaded this zone of semi-permanent water are festooned with a wide variety of organisms including large orange sponges (Tedania ignis) and numerous different species of algae.

Top: The variety of organisms growing on the permanently submerged parts of prop-rooted mangrove roots provide the basis for a miniature ecosystem to flourish. Here a foureye butterflyfish (Chaetodon capistratus) – *an animal more usually associated with coral reefs – plucks small organisms from the surface of a fire sponge* (Tedania ignis) *attached to a red mangrove root.*

Bottom: Where mangroves border sea grass one can find a considerable overlap of species between the two. This spectacular dragonet from Papua New Guinea (Dactylopus dactylopus) *normally hides in sea grass where it lowers its fins and looks like a bit of weed. When alarmed it will erect all its fins in this remarkable display. The long, plume-like extension of the anterior part of the dorsal fin identifies this as a male.*

Amphibians are rare in mangroves; their need for fresh water and their tendency to dehydration make this an unsuitable habitat. This tree frog (Hyla sp.) was in the back edge of a Bimini, Bahamas, mangal.

trough, which we may call the seaward trough zone. The submerged parts of the roots become a valuable surface for organisms to attach to. In clear water the roots can be covered in algae. Oysters, barnacles, mussels and marine snails are also common on these submerged roots. Sponges of impressive size can attach too, as can the similar-looking but much more advanced tunicates. Cnidarians may be represented by hydroids and anemones. Bristleworms and fanworms are also encountered. Given the wealth of organisms that can attach to or live on the submerged parts of the prop roots (or attach to the animals attached to the roots), it is hardly surprising that a wide variety of fishes will enter the region to feed, and will themselves attract predators. Again this narrow, shallow area offers relative safety compared to the open sea and various juvenile fishes and crustaceans hide here. Where there is a virtually seamless overlap with neighbouring sea grass or coral communities a very wide range of creatures can be encountered.

Amphibians and reptiles

Amphibians are rarely encountered in mangals. I have seen a few tree frogs (*Hyla* sp.) hiding from the heat under the leaves of buttonwood trees in the landward zone of a Bimini mangal, but the majority of amphibians need fresh water in which to deposit their eggs. The lack of fresh water in most mangroves and the tendency of amphibians to dehydrate in salty conditions virtually banishes them from the mangal.

Reptiles however are more successful. Sea snakes enter the mangal at high tide to feed. Varieties of sea

This species of sea snake (Hydrelaps darwiniensis), seen here in Darwin Harbour, is adapted to hunt on the mangrove mudflats at low tide. It readily enters burrows in search of mudskippers and presumably crabs as well.

snake such as *Hydrelaps darwiniensis* are specially adapted to hunt on mudflats at low tide, and are sometimes seen on Northern Territory mudflats poking their heads down mudskipper and fiddler crab burrows in search of a meal. Many species of terrestrial snake probably enter the mangroves via the canopy when it is contiguous with terrestrial forest.

Lizards and skinks are also commonly seen in mangrove trees, although few are thought to be dependent on mangroves for their survival. An exception is the mangrove monitor lizard (*Varanus indicus*), which can reach at least 1 m in

length and readily swims in sea water. In the mangals of tropical Australia this varanid typically lives in a hollow branch of a dead grey mangrove tree and feeds on small crabs and fishes. It takes some skill to spot one of these secretive animals. Russell Hanley says that he often spots mangrove monitor lizards because he can 'feel' them looking at him. Once when we were in a mangal on the Adelaide River he turned around and stared upwards into the canopy.

'What have you seen?' I asked.

'Nothing.'

'Then what are you staring at?'

'A mangrove monitor lizard.'

'But I thought you hadn't seen anything.'

'I haven't.'

'Then what are you looking at?'

'A mangrove monitor lizard.'

The conversation could have gone on for ever but eventually Russell admitted that he could not see what he was looking at, although he knew it was up there somewhere. I wondered if he had spent too long in the tropics but later that afternoon he

The mangrove monitor lizard (Varanus indicus) *is especially adapted to mangrove swamps. In Northern Australia it tends to be found (as in this photograph) in areas with tall grey mangrove trees. It builds its nest in the hollows of such trees.*

suddenly cut the motor of our boat and pointed to the endless, indistinguishable phalanx of mangrove trees.

'What is it?' I asked.

'Another mangrove monitor lizard.'

'Can you see it or just feel it?' I asked suspiciously.

'Yes, there, plain as day! Out on the branch of that *Camptostemon schultzii*. Here's your chance to get a picture. See it?'

All I could see was an endless wall of similar-looking mangrove trees.

'Er, which *Campt...Campt...*Which tree exactly?'

'There is only one *Camptostemon*. Quick, get your camera ready. I'll paddle the boat towards it.'

We set off towards the wall of trees.

'You *must* be able to see the monitor lizard now,' Russell said.

'Aha!' I said triumphantly.

'You've seen the monitor lizard?'

'No, but I think I've spotted the tree.'

'There. On that branch. As clear as day.'

'Oh...Right I think I see it now.'

The fates had snubbed me: I had silently mocked Russell's ability to look at a mangrove monitor lizard that he could not see; now I was having to pretend to see one that I probably could not have spotted even if it jumped out of the tree and landed on my head.

The crocodilians are the most infamous members of the mangrove reptile fauna. While crocodilians are not dependent on salt water habitats they are able to live there because they possess salt glands: when feeding a crocodile can swallow a large amount of salt water. The salt is excreted by the glands. Furthermore the two truly estuarine crocodiles – the 'salty' or estuarine crocodile of the Indo-Pacific (*Crocodylus porosus*), and the American crocodile (*Crocodylus acutus*) avoid drinking when in an estuarine habitat. The thickness of their skin also acts as an impermeable barrier against dehydration.

Their ability to live in mangrove swamps means that they are in a food-rich habitat filled with crabs, shrimps, fishes, the occasional mammal and numerous birds: a wide variety of food items of different sizes is available for a growing crocodile. Moreover, being cold blooded, they need to eat far less food than a correspondingly sized warm-blooded predator – perhaps as little as a tenth. This means that in areas where they have not been disturbed or poached, they can both be extremely common and very large.

The American crocodile (*Crocodylus acutus*) is in danger of extinction over much of its range as a result of hunting and the disturbance or destruction of its habitats. The steady reduction of mangrove areas has contributed dramatically to the reduction in its numbers. Other, smaller crocodilians can be found in New World mangals. The common cayman (*Caiman crocodilus*) seasonally enters the Caroni Swamp of Trinidad. It is not known to what extent the American crocodile displaces other crocodilians

*The common cayman (*Caiman crocodilus) *is the most adaptable species of cayman and is found in a wide range of habitats. This individual was in a rarely visited corner of Caroni Swamp, Trinidad with another cayman in the same pool of mangrove-fringed water. Common caymans enter Caroni Swamp to breed and this was presumably a breeding pair.*

Johnston's crocodile (Crocodylus johnstoni) is found only in tropical Australia. It is normally confined to fresh water – hence its other common name, the river crocodile. However, this species will eventually enter estuarine (and therefore mangrove) habitats if the estuarine crocodile (Crocodylus porosus) is removed. The 'freshy' as this species is locally nicknamed, is rarely a threat to man unless molested. It feeds on a wide variety of prey and the narrow snout is presumably an adaptation for feeding on small prey.

from the mangal areas in which it occurs. It is tempting to draw parallels with the estuarine crocodile of the Indo-Pacific, which will exist with other crocodilians in river systems, but is thought to competitively exclude them from estuarine areas. This exclusion is probably because although other members of the genus also possess salt glands they are not as good at excreting salt – which requires a lot of energy – and catching enough food. For example in the fresh water reaches of the Northern Territory rivers the smaller Johnston's crocodile (*Crocodylus johnstoni*) can be found with estuarine crocodiles. However in mangrove estuaries

The estuarine crocodile (Crocodylus porosus) occupies a wide range of low-lying terrestrial habitats: fresh water swamps and ponds, rivers, tidal creeks and mangrove forests. It is quite capable of surviving in sea water and will embark on considerable journeys across the sea in search of new habitats.

only the estuarine crocodile occurs. When its numbers were dramatically reduced by excessive hunting in northern Australia it was found that the smaller 'freshy' crocodile made considerable advances into estuarine habitats. Now that the 'salty' is protected and its numbers have more than recovered, the 'freshy' is no longer found in coastal areas. The inland penetration of the estuarine crocodile is limited by the fact that it builds its nests in low-lying floodplain areas. It is well able to swim in the open ocean and covers considerable distances in its search for new habitats to colonise. Indeed some individuals have been found with pelagic barnacles attached to them!

The estuarine crocodile has an enormous distribution – from India and Sri Lanka through the tropical belt of Asia, Indonesia, Malaysia and the Philippines, through New Guinea and northern Australia to as far east as Fiji, but habitat destruction and local hunting (for the protection of the people or the skin of the animal) have reduced its numbers in many areas.

As is so often the case with extremely large, extremely powerful and extremely dangerous predators, it is very difficult to record this animal's maximum size accurately. We have an instinctive desire to exaggerate the dimensions of the creatures we fear most, and the largest crocodiles are most likely to be in the remotest locations, where they have been allowed to grow and grow unmolested. There are thus rarely more than anecdotal size estimates of the claimed giants.

Tex Boneham is a long-retired crocodile hunter from Wyndham in Western Australia who has seen and shot hundreds of estuarine crocodiles. On one outing into the outback mangal he and a colleague came across what looked, from a distance, like a group of large estuarine crocodiles basking on a mudflat. His colleague took aim and shot and killed the nearest one. The crocodiles behind the dead one immediately started to slither into the muddy water to escape. Tex's colleague fired again and wounded one of them. Then they realised that they were not looking at several crocodiles but at one huge one. It slithered into the creek and vanished. Tex, who had been estimating the size of crocodiles for many years, rated this one at about 7.6 m, the largest he has ever seen.

Greg Harman is an adventurer who takes people on wildlife safaris in his little open boat along the Northern Territory's massive and remote Ord River. He took me out to photograph the wildlife for a couple of days, and we saw numerous 'freshy' and 'salty' crocodiles. Although Greg treated them with respect, he was quite willing to bring the boat up close so that I could get my pictures. On one occasion a medium-sized crocodile was lying on the mud of the river's edge and I asked Greg to take the boat as close as he could get to it. We went about halfway and then seemed to stop. I asked him to get closer still. He moved the boat a fraction closer but then would not take it nearer. 'I don't want to get any closer,' he said simply. 'There's a big croc that lives between us and that one. You hardly ever see him but he's down there.'

He had never shown any fear of the crocodiles before so I was a little surprised.

'How big?' I asked.

'I would guess 25 ft (7.6 m). He only comes out in a cold spell. Then you see him on the mudbank warming himself in the sun. Now it's too hot. He's on the bottom.'

We were in a very small boat. I decided that I was not as interested in photographing the visible crocodile as I had thought and decided to serendipitously spot something in the opposite direction.

'What was that on the other side of the river?' I asked.

'I don't know.' Said Greg. 'Let's go and find out.'

The estuarine crocodile is an apex predator of formidable dimensions and cunning. Its favourite method of hunting is by ambush. It is often said in the

Northern Territory coastal areas that one should never fish or wash up in the same spot more than once; there may be a crocodile nearby that notes one's habits. The next time one returns to the same spot the crocodile may be waiting.

In some parts of the world people who live at a subsistence level have no choice but to risk their lives when hunting and fishing where estuarine crocodiles occur. However there are many remote areas where people rarely venture. Tragedies can occur in these parts when inexperienced visitors go for a swim without the slightest idea of the risk they are taking. On the other hand the protection of the northern Australian crocodile population has also meant a considerable boost to the tourist industry: people will pay to see these massive animals in the wild.

Estuarine crocodile numbers can be controlled in certain areas; for instance Darwin Harbour has numerous crocodile traps – baited cages – which catch ingressing crocodiles so that they can be removed from the area. Nevertheless, given their ability to occur in enormous numbers and also their habit of travelling great distances, it is clear that if the population of the animal is allowed to reach a stable level then the risk to humans will remain.

Do estuarine crocodiles offer any benefits to the habitats in which they occur? (They use the creeks and mud flats of the mangrove, but they do not fight their way into the densely forested areas.) Any violent activity, such as hunting or fighting a rival, which stirs up the mud will certainly contribute to the aeration of the mud as well as bringing valuable, buried nutrients to the surface. Moreover, the organic material comprising the prey items eaten by the crocodile will be recycled in the mangal after the crocodile has excreted it. A crocodile which hunts transient visitors to the mangal such as tidally-dependent fishes or visiting birds, will be adding to (or helping to balance) the amount of organic carbon in the system. Furthermore a crocodile that dies in the mangal will quickly be consumed by scavengers and its carbon content returned to the ecosystem. I remember seeing a goat that had died on the edge of the Kilgwyn Swamp in Tobago. Within a few days it was a seething mass of maggots. Numerous crabs had arrived from the mangrove muds to get their share of the meal. Although the maggot eggs had been laid by terrestrial flies, half of the goat was submerged by the high tide and the maggots in this area were quickly washed off the carcass. The splashes in the water announced that there were plenty of feeding fishes to mop them up. In the same way mangroves that have crocodiles in them may have an advantage over those that do not as the crocodiles play an important part of the carbon recycling of the system both when living and when dead.

Fishes

One normally thinks of fishes as living in the water, so a trip to an Indo-Pacific mangrove swamp at low tide can be a surprise: drab little fishes of about 5 - 10 cm in length wriggle in an ungainly way across the mud. The movement has been described as 'crutching' as it well resembles the awkward movement of a person using crutches. These fishes may seem easy to approach but their bulging, high-mounted, swivelling eyes give them excellent vision and they will either vanish down a nearby burrow at your approach or, by curving their bodies into an arc and using their tails like a spring, skip quickly away over the mud. This gives them their common name: mudskippers. A close examination of the mangrove tree roots may also reveal their presence; they can climb trees because of their strong pectoral fins (equivalent to our arms). Furthermore the pelvic fins (on the underside of the animal below the pectoral fins) are fused into a suction cup that can be used to attach the animal to a suitable surface.

It is suspected that there is a zonation of mudskippers within the mangal: species with completely fused pelvic fins are found at the seaward edge, while other very similar species with less completely fused pelvic fins are found in less frequently inundated parts of the mangal. The ability rapidly to climb an object out of the reach of the incoming tide is crucial to these amphibious fishes; if one is kept submerged for too long it will drown. The reason for this is that it breathes a soup of air-bearing water. Before one sets out across the mud it will close its gills and take in a mouthful of water, which makes the head suddenly look enlarged. Within the water-sloshed gill-chamber will be a little trapped air from which the fish can extract oxygen while the gills are kept moist by the

*A pair of mudskippers (*Periophthalmus argentilineatus*) pose in a mating burrow. Because bony fishes use external fertilisation the use of a burrow (where the gametes are held together in a small area) increases the number of eggs fertilised. They are then washed out to sea by the incoming tide.*

water. Every once in a while the mudskipper will dip its head into a pool of water to replenish its gill chamber supply. The skin of this variety of mudskipper is well supplied with blood vessels. It has been suggested that these are supplementary avenues of gas exchange.

The mudskippers spotted amongst the mangrove trees usually belong to the genus *Periophthalmus*. They are carnivorous and feed on insects, worms and small crustaceans. However there are other species in other parts of the mangal habitat.

Immediately in front of the mangal there may be an extensive area of soft

The blue-spotted mudskipper (Boleophthalmus caeruleomaculatus) is a shy species of the intertidal mudflats in front of Northern Territory mangals. This one is scurrying out of a pool of water and simultaneously raising its dorsal fins.

mud sloping gently down to the low tide mark, which may be rich in other varieties of mudskipper. Because of their timidity it is difficult to approach the mudskippers without all dashing down into their burrows and staying there for a considerable amount of time. A pair of binoculars is very useful so that one can position oneself at the seaward edge of the mangal, hide behind a tree and then observe them without disturbing them.

The mudskippers most likely to be spotted in the soft mud in front of a Northern Territory mangal are instantly differentiated from their cousins within the mangal because of their larger and more thick-set bodies. The most common ones, which can be seen in enormous numbers on the exposed mudflats, belong to the genus *Boleophthalmus*. They are normally up to about 15 cm long and their technique of filling the gill cavity with water gives their head a bulbous appearance. They are less amphibious than their *Periophthalmus* neighbours as they do not seek air when the tide rises, but allow themselves to be submerged. Indeed the structure of their gills is typical of the group of fishes to which all mudskippers belong (the gobies, Gobiidae), whereas the gills of *Periophthalmus* are reduced in size.

The most commonly seen species of *Boleophthalmus* in front of the Northern Territory mangal is *B. birdsongi*, which is identified by the dark strip running the length of the body. Also present is the blue-spotted *B. caeruleomaculatus*, which may undergo radical population changes: on one visit to a certain mudflat it was about a quarter as common as *B. birdsongi*. On another visit a few months later, it had virtually vanished. (This may be because it does not regularly emerge from its burrow!)

The *Boleophthalmus* mudskippers are highly active feeders, sweeping their heads across the mud and gobbling down a wide range of tiny plants and animals. But they are of special interest because of their dramatic behaviour. A male

Top left: A mudskipper (Boleophthalmus birdsongi) approaches another and opens its mouth in a threat display. Top right: The second mudskipper responds by opening its mouth and facing its opponent. Bottom left: After a few seconds the smaller mudskipper backs down and shuts its mouth. Bottom right: The first mudskipper raises its fins in a victory signal.

will build a nest (a hole in the mud) several centimetres in diameter. This will naturally collect sea water and soon be flooded. The male then tries to attract a female. In at least one variety of *Boleophthalmus* the male builds a low mud wall around his nest so that the mudflat becomes covered in these little structures. The males then leap above the height of the wall. Presumably a female will be most impressed with the highest-jumping (i.e. the fittest) male - he will certainly be the most conspicuous!

Once a female is attracted into a male's territory he will entice her to the burrow entrance. There can be several parts to the attraction process. A brightly coloured dorsal fin that is raised and lowered is another feature of the male mudskipper's behavioural repertoire. Again, the nature of the male's leap varies between species. In one variety of mudskipper observed on Singapore mudflats, the males perform three leaps, the last little more than a shrug. When a female nears his nest the male uses his final, devastatingly seductive ploy: he puts his head in the water and blows bubbles. The female enters the nest burrow and mating takes place. Because most fishes employ external fertilisation (eggs and sperm come into contact outside the female's body) there is usually a great wastage, as many sex cells never even meet. So the fact that all the mudskipper's eggs and sperm are held together in the small pool of water in the nest burrow means that the number of eggs fertilised is relatively high.

Males of the genus *Boleophthalmus* are fiercely territorial and exclude other males from their territory by what is termed mouth-fighting. Two males will freeze head-on with their jaws agape as if poised to bite the opponent. They can remain frozen in this pose for several seconds. Sometimes they will also start pushing at each other although actual biting appears to be rare.

Several other groups of mudskipper can be found in the environ of the mangal. The largest is the predatory *Periophthalmodon schlosseri*, which can reach

25 cm and is typically found in the soft mud in the upper reaches of tidal creeks. It builds its burrow in the steep walls of the creek banks. It has two rows of teeth (the other mudskippers only have one) and is the super-predator of the mudskipper world, eating other mudskippers as well as insects and crabs.

Where the mud is very wet one can encounter the bearded mudskipper (*Scartelaos histophorus*), which is easily identified by its long, thin body (it can reach 15 cm in length) and also by its characteristic feeding method: it sweeps its head from side to side. The frilly excrescences underneath are used as a filter for trapping food items.

*This mudskipper (*Scartelaos histophorus*) is found in the semi-liquid mud in front of Northern Territory mangals. It feeds by raking its head back and forth across the mud surface. When alarmed it will dive into the mud.*

The density of mudskippers on food-rich mudflats can be remarkably high and complex forms of aggressive behaviour have evolved which establish dominance not only within a species but also between species. The raising and lowering of the dorsal fin ('flag waving') has already been mentioned in the context of mating but flag waving is also employed aggressively. Mudskippers can also change colour in order to intimidate an opponent. There is some debate as to whether they are strictly territorial: some scientists think that territoriality only develops in response to nest building while others think that a travelling mudskipper will take its 'territory' with it – in other words aggressively challenge other mudskippers (and even, on occasion, crabs) which venture too close.

There is a clear-cut partitioning of habitat by different species of mudskipper: one species will prefer the soft mud on the edge of creek banks, another will inhabit more exposed areas of liquid mud. Where two species co-exist, a dominance hierarchy is usually found.

There are other species of mudskipper which live below the surface of the more liquid mud, but they are rarely if ever seen because of their secretive habits.

In the mangroves of Central America there is an interesting group of fishes known locally (if wrongly) as mudskippers. These are fishes of the genus *Anableps*. They are characterised by their extraordinary eye design – indeed another common name is the four-eye fish. The eyes are mounted bulbously on the top of the head and divided in two so that the upper half can focus in air and the lower half in water. The fish swims along the surface with the eyes held half in

The four-eye fish (Anableps sp.) of Central American rivers and creeks has protruding eyes that are held half in and half out of the water. Different lens structures in the different parts of the eye allow the fish to focus simultaneously both above and below the water.

and half out of the water so that it can spot the approach of danger from both domains. Unlike the mudskipper, *Anableps* is entirely aquatic in its habits, and is an example of a fish adapted to shallow, muddy water – which need not include mangrove areas. Given the range of habitats that can be occupied by mangroves (from the seaward edge of a coastline without fresh-water input to the banks of a river above the influence of salt water), a study of all the fishes found within them would be vast and range from exclusively fresh-water to exclusively marine species. A scientist who collected the fishes from 14 different mangrove swamps – two in northern Australia, six in Irian Jaya and six in Papua New Guinea, counted 204 species from 58 families. What would be interesting to establish is whether there is a fish population that can be considered to be typical of a mangrove community as opposed to being merely typical of the inshore, estuarine location. Given that different mangals can be comprised of very different species of tree – one

Aggregations of grey snappers (Lutjanus griseus) *are common in the seaward trough zone of Bahamian mangroves. Indeed the regularity of their occurrence in this habitat means that they are often referred to as mangrove snappers.*

Mangrove cardinalfishes (Sphaeramia orbicularis) shelter under the prop roots of a stilt-rooted mangrove in Papua New Guinea. This species of cardinalfish is heavily dependent on the shelter provided by mangroves in general and prop-rooted mangroves in particular. Note the bulge under the mouth of the central fish. This is a male and, typically of cardinalfishes, he is incubating eggs in his mouth cavity.

mangal may be entirely composed of the mangrove palm for example, while another has a mangrove apple zone leading back to a knee-rooted mangrove zone and so on – one must further ask to what extent the fish community is a response to the floral composition of the community in question.

The organic richness of the waters coming out of mangrove swamps will obviously attract fishes to the area. Indeed it has been suggested that as many as 90 per cent of tropical marine inshore fishes spend at least part of their life cycle within the mangrove swamp. Not only is it a food-rich medium, but the shapes of the submerged trees provide protection from predators. A dense area of pneumatophores or the buttress roots of prop-rooted mangroves can provide an invaluable hiding place for a small fish. For example the cardinalfishes (Apogonidae) are nocturnal planktivores, and during the day they seek shelter - on coral reefs they are to be found hiding in coral crevices. The seaward trough zone of an Indo-Pacific mangal which is filled with prop roots from the local mangroves provides similar shelter. And indeed, in my excursions into Papuan mangals I always found a particular species that might be called the mangrove cardinalfish (*Sphaeramia orbicularis*) in amongst the prop roots wherever a seaward edge of *Rhizophora* was encountered. Indeed even individual *Rhizophora* trees that were not part of a general mangrove community still had their mangrove cardinalfishes sheltering inside them. Given that I did not see this fish in other habitats and that when it was encountered in mangroves it was always in the *Rhizophora* zone (and nearly always sheltering within the root system of a prop-rooted mangrove), this suggests considerable habitat specificity.

The archerfishes (*Toxotes* spp.) are common in Indo-Pacific mangroves and worthy of mention because of one of their astonishing feeding methods. While they will feed on small marine creatures this is not their main claim to fame. An archerfish usually swims just below the surface. It has excellent binocular vision and will occasionally leap out of the water to grab an animal such as a spider suspended in an overhanging web. This is no small feat, as it must correct for the distorting affect of light rays passing

*This archerfish (*Toxotes jaculator*) was photographed in a Papuan mangal. It derives its common name (and fame) from the fact that it can squirt droplets of water up through the surface and shoot down insects which it then devours.*

Below: This seahorse (Hippocampus kuda) *is using its prehensile tail to cling on to a root just off the bottom. Its yellow colour blends with the leaf litter around it.*
Left: This seahorse clinging higher up on a prop root has assumed a darker coloration to blend into its more sombre setting.

from air to water: they bend, so that an object in the air viewed through water is not where it appears to be. Far more remarkable, however, is the archerfish's ability to shoot droplets of water at insects on the overhanging trees. There is a tube-like groove in the roof of the archerfish's mouth, and rapid contraction of the gills forces water along it and out through the mouth, with the tongue controlling the rate of fire. It is reputed to be deadly accurate at up to 1 m and, after a few preliminary range-finding salvoes, effective at up to 3 m.

In many parts of the world mangroves line the coast and are anchored on a substrate mostly composed of coral rock or sand. In these areas the visibility can be quite good and snorkelling in such a mangal (having first established that there are no dangers such as the box jellyfish or crocodiles) can be fascinating. Where the mangroves meet sea grass beds or areas of living coral one can usually find creatures more typical of these other habitats which have made an incursion into the mangroves. Many fishes are able to change colour, assuming a different coloration according to their new habitat. Indeed the adaptations can be more specific still: one species of seahorse (*Hippocampus kuda*) is common on the inshore fringe of sea grass in the shallow waters of Papua New Guinea. I remember encountering two in the mangrove trough zone of a coastal fringe of mangroves in the Milne Bay district of Papua. The first was bright yellow in colour and blended with the fallen yellow leaves around it on the sea floor. The second was attached halfway up a stilt-rooted mangrove's prop root and had assumed

This little frilly blenny (Petroscirtes mitratus) is well camouflaged in the deep shade of algal-covered mangrove roots in a Papuan mangal. Presumably the fish feeds on the algae attached to the roots.

Top: The filefish Acreichthys tomentosus *is common in inshore seagrass areas in Papua New Guinea. When alarmed, it pretends to be a dead leaf. This filefish is also found amongst mangrove roots where its mimicry is put to good effect. There are actually two filefishes in this picture.*

*Bottom: Pistol shrimps (*Alpheus *spp.) have one greatly enlarged claw which is used to make gun-shot noises. They occur in mangroves and can often be heard when the tide has gone out. In clearer water they often share their burrow with any one of numerous species of goby. The shrimp has poor eyesight and the goby keeps watch for danger while the shrimp provides the goby with a home. Here, in a Papuan mangal, an alpheid shrimp shares its burrow with the goby* Cryptocentrus leptacanthus.

a brown colour to blend with this more sombre setting. In fact I almost failed to spot it as I was trying to photograph a pair of archerfishes which were flitting nervously back and forth in the same place. Gradually, as my eyes became adjusted to the gloom, I began to pick out numerous other inconspicuous fishes that were blending with the mangrove world: a frilly blenny (*Petroscirtes mitratus*) that was virtually indistinguishable from the algae-covered prop roots on which it fed; a filefish (*Acreichthys tomentosus*) that pretended it was just another leaf on the bottom. Where the seaward trough zone is clear enough, mutually

*This beautiful goby (*Amblygobius phalaena*) has made its home in a trough of permanent water below some mangroves in Papua New Guinea.*

advantageous relationships normally associated with clear waters can be established. For example in the muddy, low-visibility waters that typify the mangal habitat one can find pistol shrimps (*Alpheus* spp.) which build burrows in the mud for shelter. However where the water is clearer pistol shrimps often share their burrows with various species of goby: the goby has far better eyesight than the shrimp and acts as its look-out. The shrimp builds the burrow in which the goby lives and in exchange the goby offers visual protection. In Papua New Guinea the seaward trough zone of the mangal often has such mutual pairings where the water is clear enough.

An underwater exploration of the mangroves must be conducted extremely slowly and patiently, not only because it is all too easy to stir up the silt and ruin the visibility, but also because so many of the underwater creatures are extremely good at avoiding being spotted. Since one does not know what to expect it is easy to overlook what is there. On one occasion in Milne Bay, Papua New Guinea,

These transparent shrimps (Periclimenes sp.) *gather in groups and set up 'cleaning stations': areas where fishes arrive to have their parasites removed from them by the shrimps. Here, such a cleaning station has been established in the mangroves of Papua New Guinea.*

I was trying to photograph a goby and a pistol shrimp which shared a burrow in the mangrove trough zone of a belt of stilt-rooted mangroves. Because of the timidity of the creatures and their tendency to scamper down into the safety of their burrow, I had to keep completely still for many minutes waiting for them to emerge. During the wait I felt a tiny tickling sensation in my right ear. I thought nothing of it, but then I felt the same sensation in my left ear. I brushed my hands across my ears but a few seconds later they started tickling again. I could not work out what was going on. Then I felt a tiny itch on the back of my hand. I looked at it and saw an almost completely transparent shrimp picking away at my skin. By gently wafting my hands across my ears I was able to reveal several other shrimps that were similarly occupied there. I then looked closely at the mangrove prop roots around me and saw that there were several dozen more of these almost invisible creatures. They belong to the genus *Periclimenes* and actually clean parasites from the skin of fishes which present themselves at their cleaning stations. These stations are well known from numerous shallow-water locations such as coral reefs and sea grass beds. I had inadvertently presented myself to be cleaned by taking up position in this location. It is interesting to note (although not, in retrospect very surprising) that cleaner shrimps have cleaning stations in the clearer parts of mangroves. The fishes in the mangroves are every bit as susceptible to parasites as others.

Birds

It has long been known to ornithologists that mangroves provide a spectacular setting for certain kinds of bird. The exposed mudflats and shallow creeks and pools of water are rich in a tremendous range of food items: worms, crabs, shrimps, molluscs, insects and fishes are all predated by birds in the mangal.

*Left: A pair of black-necked storks (*Ephippiorhynchus asiaticus*) feed on mudflats in front of Darwin Harbour mangroves. The female has the orange eye. These impressive birds reach a height of 120 cm when mature. They are expert fish-catchers, using their sharp bill to stab at fish in pools of water.*
*Above: Certain species of egret and heron will take over specific trees for nesting purposes. This pair of cattle egret chicks (*Bubulcus ibis*) was photographed in a Caroni Swamp, Trinidad, red mangrove tree covered in nesting egrets.*

The mudflats exposed by the falling tide may become covered in various types of shore and wading bird. Indeed a quiet observer in a mangal can expect to see examples from an impressive range feeding along the fringe of the retreating tide, including herons, ibises, plovers, curlews, whimbrels, dowitchers, sandpipers, seagulls and kingfishers.

Some birds nest in the mangroves. For instance the cattle egret (*Bubulcus ibis*), a relative newcomer to Trinidad and Tobago (the first record of the species is

*This nesting tricoloured heron (*Egretta tricolor*) shares its nesting tree in Buccoo Swamp, Tobago, with numerous other tricoloured herons as well as a great many cattle egrets.*

in 1951) forms noisy and densely packed nesting colonies in the mangroves of Caroni Swamp in Trinidad and Buccoo Swamp in Tobago. In Caroni Swamp it nests with snowy egrets (*Egretta thula*). In Buccoo Swamp it nests with tricoloured herons (*Egretta tricolor*). Other birds such as the scarlet ibis (*Eudocimus ruber*) gather to roost by the hundred in specific trees in Caroni Swamp at dusk. The large amounts of guano regularly excreted by the birds in these trees provide an important fertiliser that is rich in nitrates and phosphates. The chosen trees can often grow taller than others in the area and produce noticeably denser foliage.

Where a species of bird regularly gathers to feed the particular area may become important for social interactions as well. For example in Caroni Swamp numerous snowy egrets can be seen feeding on certain areas of exposed mud. During the breeding season, when they assume their spectacularly delicate plumage, the males are often to be seen performing their elaborate courting rituals. Elements of the display include stretching the neck and flapping the wings as well as flying in tight circles and taking

While most scarlet ibises (Eudocimus ruber) *feed in the impenetrable depths of Caroni Swamp, Trinidad, a few feed in the brackish areas of needle grass bordering the swamp.*

off vertically, flapping a couple of metres aloft only to return clumsily to the ground.

The spectacle for which Caroni Swamp is world famous, however, is the roosting of the scarlet ibises. During the day these startlingly coloured birds are dispersed throughout the swamp. Some will be seen perched high on the branches of red mangrove trees. A few others might be seen feeding at low tide on the mudflats. The vast majority, however, are gathered deep in the impenetrable depths of the swamp feeding amongst the red mangrove roots. The scarlet ibis is a wary bird: it has been regularly poached (blasted with shotguns) despite being Trinidad and Tobago's national bird and supposedly protected. Every few hours in Caroni Swamp one can expect to hear the din of a shotgun signalling the activities of a poacher. The result is that the birds no longer nest in Caroni Swamp but fly to Venezuela in order to breed. It is suspected that their numbers are steadily falling in Trinidad as fewer and fewer birds return to Caroni Swamp after the nesting season. It will be a sad day if the country loses its national bird because of the lack of sufficient controls against poaching.

Nevertheless the sight that greets the visitor to Caroni Swamp at dusk is still one of those magical events that people travel great distances to observe. Winston Nanan, a local ornithologist and natural history expert, has several small boats and he and his guides take birdwatchers to within a few hundred metres of the appropriate trees just before the sun sets. If the tide is out then numerous snowy egrets might be visible feeding on the mud in front of the red mangrove trees. Then they leave the mud and fly into various trees in order to roost for the night. The trees seem to sparkle with numerous white

In the last glint of daylight scarlet ibises fly towards their roosting tree.

The scarlet ibises are the jewels in the crown of Caroni Swamp. At dusk they gather by the hundred to roost in certain trees.

dots in the failing light. Shortly thereafter (and just after the trees are cast into shadow and the conditions for photography have deteriorated), squadrons of scarlet ibises swoop in from other parts of the swamp and begin to settle in their favourite trees. Within a few minutes several hundred scarlet ibises can be seen roosting in a single tree and their points of red fire give the trees a mysterious beauty. Unfortunately it is by now almost dark and photography is virtually impossible. I tried to photograph them on several evenings but they only arrived in their roosting tree when it was cast into deep shade. Winston and I went out at dawn on several occasions to try to photograph the spectacle before the birds left for the day. We would slip quietly into position, and the scarlet ibises could be clearly seen in their hundreds in the trees as the light grew stronger. But again, as if deliberately to frustrate my attempts, they would all fly off a few moments before the sun lit up their trees.

As with so much of the mangrove fauna it is difficult to say exactly which birds are regular and which are occasional visitors to the mangroves, and which are only found there. Another question is whether the birds found in the vicinity of mangroves are in fact dependent on the food-bearing mudflats of the area rather than the mangroves themselves. Certainly most if not all visiting shore birds would fall into this latter category, and of course the mangrove trees can provide convenient roosts for birds when the tide is in. In Australia the lesser noddy (*Anous tenuirostris*) tends only to breed in the mangrove trees of Western Australia's Albrolhos Islands.

Surveys in Australia's mangroves have revealed over 200 species of birds. While it is claimed that the avifauna of Australian mangals is substantially richer

*Opposite page: An immature little blue heron (*Egretta caerulea*) stands on a red mangrove prop root in Caroni Swamp.*

than that of other parts of the world it is more likely that people have simply not looked as hard in other places. For example the vast and towering mangrove forests of Papua's Fly River region, which is fairly close, are surely rich in bird species, although only three species are claimed to be common. Given the huge height of the trees and the density of the foliage, bird spotting there would be immensely difficult. An equivalently bathetic figure for bird numbers has been claimed for Bornean and Malaysian mangals, doubtless for the same reason. Of those counted in Australian mangals only about 14 species are virtually restricted to the mangroves. Seasonal fluctuations in mangrove avifauna can occur due to migration patterns of birds and also because flowering mangroves attract either nectarivorous birds or birds that feed on nectarivorous insects.

Certain birds have been shown to require specific areas within the mangal. For example the mangrove robin tends to be found in prop-rooted mangrove areas: it perches on the roots and also needs the rich supply of insects the area contains. Several species of kingfisher can be found within Northern Territory mangals, although there seems to be little intermingling of species within the same mangal. The mangrove kingfisher builds its nest in the hollows of the grey mangrove tree. In the Northern Territory it is very common in the Ludmilla Creek mangal. This was the point at which Cyclone Tracy came ashore in 1974. Many of the grey mangrove trees are damaged or dead and this might explain the unusually high concentration of mangrove kingfishers here.

The red-headed honeyeater, although small (up to about 13 cm), is a highly conspicuous member of the Northern Territory mangal avifauna. The male is mostly darkly coloured but with a scarlet head; it is remarkably active and pugnacious and seems to spend most of its time chasing other similarly-sized birds through the canopy. It will settle on a branch and gaze about for as long as it takes to spot another bird that is about its own size (including

Table 6. *Birds heavily or exclusively dependent on Australian mangroves* (Adapted from Johnstone (1990).

Great-billed heron	*Ardea sumatrana*
Mangrove (green-backed) heron	*Butorides striatus*
Grey goshawk	*Accipiter novaehollandiae*
Chestnut rail	*Eulabeornis castaneoventris*
Bar-shouldered dove	*Geopelia humeralis*
Little bronze-cuckoo	*Chrysococcyx malayanus*
Mangrove kingfisher	*Halcyon chloris*
Mangrove gerygone	*Gerygone laevigaster*
Lemon-breasted flycatcher	*Microeca flavigaster*
Mangrove robin	*Eopsaltria pulverulenta*
Mangrove golden whistler	*Pachycephala melanura*
White-breasted whistler	*Pachycephala lanioides*
Little shrike-thrush	*Colluricincla megarhyncha parvula*
Mangrove grey fantail	*Rhipidura fuliginosa phasiana*
Northern fantail	*Rhipidura rufiventris*
Rufous fantail	*Rhipidura rufifrons*
Broad-billed flycatcher	*Myiagra ruficollis*
Shining flycatcher	*Myiagra alecto*
Yellow white-eye	*Zosterops lutea*
Red-headed honeyeater	*Myzomela erythrocephala*
Varied honeyeater	*Lichenostomus versicolor*
Mangrove honeyeater	*Lichenostomus fasciogularis*
White-breasted woodswallow	*Artamus leucorhynchus*
Black butcherbird	*Cracticus quoyi*

*This is a male red-headed honeyeater (*Myzomela erythrocephala) *from a Northern Territory mangal. The female is more drably coloured. The male of the species seems to spend much of its time manically chasing other similarly-sized birds through the mangal.*

Mangrove kingfishers are unusually common in the mangroves of Ludmilla Creek in Darwin Harbour. This may be due to cyclone damage; they nest in the hollows of grey mangrove trees and there are many grey mangroves in Ludmilla Creek damaged by Cyclone Tracy, which struck in 1974.

members of its own species) and then it will dash off after its victim. It seems to save its most maniacal demonstrations of truculence for its encounters with the long-suffering brown honeyeater (*Lichmera indistincta*), which it chases indefatigably through the foliage. The two birds are similar in size and presumably feed on the same food – chiefly the nectar from the flowers of mangrove trees. Perhaps the interspecific aggressiveness of the red-headed honeyeater is a result of the competition between the two species for virtually the same niche.

The chestnut rail (Eulabeornis castaneoventris) *is a rarely seen bird of northern Australian mangrove forests. However the secrecy may be due to the fact that people rarely sit quietly and wait long enough in the mangal. The chestnut rail is adept at plucking crabs out of their burrows.*

The chestnut rail is rarely seen because it favours dense, undisturbed mangrove forests. Indeed very little is known about this bird precisely because it is so rarely encountered and supposedly extremely shy. Nevertheless it has been seen regularly in certain Northern Territory mangals by Russell Hanley and his team. On one occasion Russell's helpers were doing a faunal survey of the substrate: recording the animals found within 1 m square transects of the muddy floor of the tidal forest. They were quietly counting the crabs and snails when a chestnut rail came along and marched through the transect: this put them in the implausible position of having to record what is meant to be an extremely secretive and rare bird as being a constituent of the transect! It was expertly darting its long bill into crab burrows and feeding on the crabs inside them. On the two occasions when I was taken to the same place, one or two chestnut rails would come wandering past after perhaps half an hour of sitting quietly. If we did not move they showed no fear but as soon as I pointed a camera in their direction they had a genius for placing themselves so that as many trees, branches and roots as possible were between me and them!

In all likelihood the chestnut rail is common enough in undisturbed mangroves. Given that a sizeable bird that actually feeds on the substrate (in other words is in the same area as human explorers) is considered rare, one wonders whether any sensible assessment can yet be made of the birds of the mangrove canopy in areas where the trees are 20 or 30 m high, ill lit and dense with foliage.

Mammals

Although mammals are often fairly common in mangals they are rarely a conspicuous component. In Australia various rodents, bandicoots and at least one species of possum are known from mangrove habitats. The crab-eating rat *Xeromys myoides* uses mangroves as primary habitat and will climb trees to avoid the incoming tide. The shade offered by the trees at the rear of the mangal is often used by various species of kangaroo and wallaby to avoid the worst heat of the day.

The proboscis monkey (*Nasalis larvatus*) of Borneo is confined to the mangroves there, and destruction of the habitat is resulting in the decline of the species. Bats will also enter the mangroves to feed on fruit and some species will roost in the trees.

The most extensive mangrove forests in the world are to be found in the Ganges Delta in the Bay of Bengal. This region, known as the Sundarbans, is virtually uninhabited. It is a refuge for the Bengal tiger (*Leo tigris*) which feeds on the numerous deer in the mangroves. This variety of the Bengal tiger grows larger than its terrestrial cousins and has a more reddish coloration. Population pressures on the peoples neighbouring the mangroves have driven them into the mangal to gather fuel wood, to fish and to collect honey. Bengal tigers regularly take people in the mangroves and are thought to stalk them and follow their boats through the swamps, and will even swim to a boat anchored in a mangrove creek and drag a person away. There are also stories of the tigers leaping from overhanging branches into boats and dashing off with

The pygmy anteater (Cyclopes didactylus) is a nocturnal feeder in Caroni Swamp, Trinidad. During the day it wraps its prehensile tail around a tree and looks like a small ball of fur. At night it feeds on termites.

their victim. Given how difficult it is for a tiger to catch something as agile and swift as a deer it seems that a human being must be a much easier option. Being largely nocturnal in their habits they are rarely seen.

The Arabian camel (*Camelus dromedarius*) will often enter mangroves in the Arabian region to feed on the vegetation. Camel owners will herd them in high numbers to feed on the mangroves and this can severely damage or kill the trees.

In New World mangroves raccoons (*Procyon* spp.) are locally common, feeding on a wide range of food items including crabs, fishes, molluscs and insects. The pygmy anteater (*Cyclopes didactylus*) is a fairly common sight in Trinidad's Caroni Swamp.

he Human Dimension

Exploitation of the mangroves

The question that has increasingly to be faced is what the effect is of human beings on the mangrove ecosystem, both locally and at the global level. The significance of the mangal both as a complex and abundant ecosystem in its own right and as a crucial organic provider for neighbouring ecosystems has long been known by scientists. Extensive removal of mangroves can damage or even devastate dependent neighbouring habitats.

The human impact on mangroves can be divided into radically different types. The first is the effect of peoples who have either lived within mangroves for several generations without destroying them or visited them to make use of their resources so that their present level of impact can continue in perpetuity. These subsistence economies are becoming increasingly rare thanks to the usually disruptive effects of economic development, as well as the rapid increase in the utilisation of resources that such development all too often

A fisherman draws a net across a creek in Caroni Swamp, Trinidad. Increasing human pressure on mangroves worldwide threatens the survival of this vital ecosystem.

entails. Nevertheless the principle of sustainability, however difficult to define and however open to debate, is basic to any conservation strategy. The premise that the less the human impact on an ecosystem the better, is axiomatic.

Humans are as entitled to exploit the mangroves as they are any other habitat, and although the impact from mangal to mangal is likely to range from the inconspicuous to the devastating, when one considers the two extremes there is no question that some types of utilisation are justified and others are not. A disturbed mangal – for example one with a proliferation of mangrove ferns owing to felling – is better than no mangal at all.

In numerous parts of the world peoples live at a subsistence level in houses and villages constructed within mangroves or close enough to make regular journeys to them. Traditional fishing methods tend to target the fishes, shrimps, shellfish and mud crabs of the mangal. Furthermore different species of tree are often felled for a very wide variety of uses – for medicinal purposes, for fishing implements, for boats, for charcoal, for house-building, for dye. Table 6 lists some examples of traditional uses to which mangrove plants are put.

Naturally enough many of the items that are of value to mangrove-dependent subsistence peoples can be of considerable commercial value and can be traded outside the immediate area. Fishes, crustaceans and shellfish can be sold to neighbouring villages, or transported to towns and cities. The peoples of the mangroves are under commercial pressure to catch more and use more efficient fishing methods that are themselves derived from developed economies. The principle of sustainability can soon be violated and the particular fishery collapse. If this only occurs locally then there is every reason to hope that the stock will eventually recover, since the animals can repopulate the mangal habitat as larvae derived from other areas.

Some mangrove timber such as that of the woody-fruited mangroves, is of considerable commercial value and has been harvested, usually with disastrous results. The damage inflicted is usually far worse than just the removal of the target trees. A path has to be made into the mangal to reach the trees and to take them out. This requires heavy machinery. The subtle variations in soil characteristics, drainage patterns and zonation are all disrupted by the dredging of deep, broad channels which totally alter the effects of the tide. Large areas of mangal thus die off in order to remove a small percentage of the trees. Furthermore it is not yet known what effect the removal of specific trees from a mangal will have on the health of the ecosystem as a whole.

Table 6. *Examples of Traditional Uses of Mangroves from around the World*

Mangrove/Mangrove associate.	Uses
River mangrove (*Aegiceras corniculatum*)	Oyster stakes. Huts. Firewood. Medicine for earache.
Pencil-rooted mangroves (*Avicennia* spp.)	Fruit is eaten (after treatment). Ash used for skin disorders and for soap. Honey collected from bees that feed on the flowers. Firewood. Trunks used for canoes, masts, furniture.
Knee-rooted mangroves (*Bruguiera* spp.)	Paddles for boats. Firewood. Structural supporting poles for houses.
Schultz's mangrove (*Camptostemon schultzii*)	Wood used for fishing floats, boats, canoes. Used medicinally to treat various skin disorders.
Freshwater mangrove (*Barringtonia racemosa*)	Fish poison.
Yellow mangrove (*Ceriops tagal*)	Various medicinal uses. Dye extracted from bark. Fishing spears and poles. Fencing.
Milky mangroves (*Excoecaria* spp.)	White latex used medicinally. Timber used for carving, fishing, utensils and firewood.

Mangrove/Mangrove associate.	Uses
Native hibiscus (*Hibiscus tiliaceus*)	Wood used for spears, fishing implements and carvings. Bark used medicinally.
Lumnitzer's mangroves (*Lumnitzera* spp.)	Durable timber used in many ways. Firewood.
Mangrove palm (*Nypa fruticans*)	Sap used to make alcohol. Fronds used for baskets, mats, thatching, cigarette wrapping. Fruit edible. Rope, brushes made from fibres.
Prop-rooted mangroves (*Rhizophora* spp.)	Charcoal. Firewood. Fishing poles. House-building. Tannin extraction (for dye). Alcohol production. Grain-sifting baskets.
Peg-rooted mangroves (*Sonneratia* spp.)	Fishing floats. Masts. Fuelwood. Construction timber.
Woody-fruited mangroves (*Xylocarpus* spp.)	Furniture. Firewood. Oil extracts. Dye extraction. Construction-grade timber.

People live in the mangroves in many parts of the tropics. This photograph was taken on a mangrove-fringed shore in Papua New Guinea. The mangroves can well survive modest human exploitation using traditional fishing techniques.

Many varieties of mangrove wood are used by subsistence peoples for charcoal production. Charcoal is the basic fuel source for cooking in many parts of the world and vast areas of terrestrial forest have been destroyed as a result. Inevitably mangrove swamps are being exploited in increasing numbers as a source of domestic charcoal. The most frequently used trees are the prop-rooted mangroves. There are many factors that make them especially good for charcoal: the tannin in the wood makes the fuel relatively smokeless and even flavours the food.

Moreover prop-rooted mangrove woods contain little moisture and a relatively high amount of combustible oils so that they are easy to ignite.

More alarming than the small-scale domestic use of charcoal is the accelerating growth in its commercial exploitation. Logging companies involved in the charcoal industry will quickly switch from terrestrial trees to mangroves when they have exhausted the supply of the former or have been prevented from taking more trees. I remember a risible explanation given by a United Kingdom spokesman for a charcoal company when asked why his company was using mangrove trees for charcoal production. He explained that by removing the mangrove trees his company was doing the rainforest trees a favour, as the mangroves were in fact choking the rainforest trees!

In 1993 the United Kingdom branch of the World Wide Fund for Nature (WWF-UK) prepared a long-needed report on the pressures facing mangrove forests around the world and I acknowledge my debt to that report (prepared by M.G. Wenban-Smith) in what follows.

From 1986 to 1990 the combined annual weight of charcoal exported from Thailand, Indonesia and Malaysia was estimated at 104,000 tonnes. It is thought that much of this was derived from mangroves, because mangroves provide quality charcoal and most of the charcoal used domestically is mangrove-derived. Alarmingly however, far more mangrove-derived charcoal is thought to be used domestically in these countries than is exported. In Thailand as much as 83 per cent of mangrove charcoal is thought to be utilised domestically. The total amount of mangrove-derived charcoal presently used by Thailand is estimated at about 200,000 tonnes per year.

Although unregulated charcoal production can clearly devastate mangrove forests, *regulated* charcoal production is not necessarily a bad thing. In certain parts of the world managed charcoal production from mangroves seems to provide the sustainable framework for a lucrative activity. One example is Malaysia's Matang Forest Reserve. This forest, comprising some 40,000 hectares, has been managed for 90 years. Trees must reach a certain size before being felled and areas are felled with a rotation period of between 30 and 40 years. A further restriction is that the maximum area of felling is 300 hectares. In 1984 this forest generated some $US 330,000 as well as providing employment for more than 1000 people. It is important that felling and transporting techniques in such forests should be designed for minimal impact, rather than the sort of obliterative methods so readily employed by multinational logging companies.

This example suggests that countries with extensive mangroves may be able to produce adequate amounts of charcoal for domestic and export use on a sustainable basis. Central to the thesis is the fact that sensitive management allows the forest to regenerate.

Mangroves have also been heavily harvested for wood chips. In the 1970s and 1980s the Japanese harvested mangroves for rayon and pulp production. Rather than limiting the areas cut to sizes in which mangroves could re-establish themselves, Japanese wood-chip production was characterised by clearing blocks of up to 4000 hectares, and this destroyed the forests. In 1976, 230,000 tonnes of mangrove wood were cut for wood chips in Sarawak, Malaysia. In Kalimantan, Indonesia, an annual production of 252,000 tonnes began in 1975 in a single forest of 85,000 hectares. It is claimed that felling activities in these areas have now been halted. Nevertheless other countries with extensive virgin mangal must be vulnerable to the attentions of the wood-chipping brigade. Irian Jaya is thought to be under consideration. Certainly those countries that are prepared to have their rainforests destroyed despite the international outcry are hardly likely to be much worried about the decimation of their mangrove forests.

The sandy mangrove (Lumnitzera racemosa) is one of the most common landward edge mangroves in Northern Territory mangals. It and its close relative L. littorea *have quality timber which makes them vulnerable to felling.*

Clearance

Another ubiquitous pressure on mangroves forests is their clearance so that the intertidal habitat can be used for prawn- or fish-farming. In Asia mud flats and local areas of mangroves have been used for such farming for hundreds of years without seriously affecting the mangrove forests. However the production of aquaculture ponds has now reached epidemic proportions in both Asia and Central America. The reasons are not difficult to understand: valuable fishes and prawns can be reared in cheap holes in the ground. Governments have been quick to subsidise aquaculture projects and huge areas of mangroves have been cleared as a result. For example in Ecuador it has been estimated that since 1970 some 35 per cent of the nation's mangrove forests have been destroyed for aquaculture production. In the Philippines fish pond areas have expanded from 88,000 hectares in the 1950's to 210,000 hectares in 1987. In the Indo-Pacific region some 800,000 hectares of mangrove forest were converted to aquaculture ponds between 1973 and 1977. In Bangladesh's Chakaria Sundarbans the influence of wealthy individuals combined with that of the fisheries department resulted in the forests losing their protected status. An estimated 30 per cent were subsequently destroyed.

It was originally thought that aquaculture was a perfect get-rich-quick industry. There are, however, considerable costs involved in the refrigerated storage and rapid transport of the fishes and prawns to urban areas. When excessive amounts of the fishes and prawns flood the market their value drops and many aquaculture

Mangrove-fringed lagoons can be nursery areas to a great variety of fishes. The lagoon of Bimini in the Bahamas is a nursery area for juvenile lemon sharks (Negaprion brevirostris). Here they are relatively safe from large predators and the mangroves offer both a wealth of fishes to feed on and also numerous hiding areas.

farms fail. In some parts of the Philippines half the aquaculture ponds have failed and been abandoned. Meanwhile, alterations to the intertidal zone produced by the aquaculture activities have exposed the mud to the atmosphere for extended periods of time. The reaction of the sulphur compounds in the mud with atmospheric oxygen produces acidification of the soil which is inimical to plant growth and has made the soil useless for recolonisation by mangroves.

The juvenile prawns and fishes that stock the aquaculture ponds are obtained from the area in which the ponds are built. This can mean an enormous number of people regularly fishing for the juvenile prawns and fishes. These juveniles not only depend on the mangroves for this stage of their development, but are predated on by estuarine fishes which are themselves indirectly dependent on the mangroves. Thus not only is the habitat destroyed but a crucial constituent of the faunal component of the region is removed as well. This decimates inshore commercial fisheries. Time and again marine biologists have insisted on the crucial part played by the mangrove habitat for inshore animal populations and time and again large-scale, short-lived aquaculture projects have been allowed to go ahead. Figures quoted (in the WWF-UK report) include the following:

- 94 per cent of the shrimps in the Gulf of Panama breed in the mangroves

- 90 per cent of the commercial fish species in the Gulf of Mexico depend on the mangroves for some part of their life cycle

- 67 per cent of Australia's commercial fish harvest depends on the estuarine mangroves

- 20,000 hectares of Nicaraguan mangrove produce 5 million kg of shrimps with a commercial value of $US 34,000,000 per year.

Many an impressive predator spends the first, vulnerable part of its life sheltering in the mangroves. This is a juvenile great barracuda (Sphyraena barracuda) from Bimini in the Bahamas.

It is depressing how often mangroves are used as a rubbish dump. Here a corner of Tobago's Kilgwyn Swamp has been turned into a dump for vehicle parts. This illustrates a widespread view about the supposed worthlessness of mangroves.

Another cause for concern is the clearing of mangroves for land reclamation for numerous uses. Mangroves tend to be found in sheltered embayments: in other words in areas that would make good harbours and they are vulnerable to removal on this score alone. The land may be reclaimed for agricultural purposes, such as the cultivation of salt-tolerant varieties of rice, but as with the aquaculture clearances, all too often the aeration of the soil leads to its acidification, owing to the sulphides it contains.

To many a developer viewing a sheltered, tropical coastline, the mangroves fringing it are a curse to be removed as quickly as possible: they block the view of the sea and they are full of biting insects. Remove the mangroves and the value of the land soars. The extensive mangrove-fringed estuary of Darwin Harbour has Australia's Northern Territory's capital on its north-eastern promontory. Much of the rest of the estuary is fringed by virgin mangroves. But for how much longer? A new harbour development has already claimed 2500 hectares. Meanwhile the plan is to increase the population of the city from the present 70,000 to 1,000,000 in the next 50 years, and presumably most of those people are going to want a view of the sea.

At present houses in Darwin are being built within $1\frac{1}{2}$ kilometres of the mangroves. Even at this distance the residents are plagued by biting midges, which can make life unbearable, so the residents complain and the government clears the neighbouring mangroves. With the mangroves cleared developers move in and build new homes on the reclaimed land. These new residents are now plagued by biting midges from the next nearest batch of mangroves and they also complain, until those mangroves are cleared and the developers move in, and so on.

Trinidad's magical Caroni Swamp is under threat from yet another direction. There are plans afoot to turn it into a recreational park! Presumably those who are considering this idea are looking for an apparently worthless resource which can be turned into something supposedly of value to the people at virtually no cost to themselves. They look at the vast, quiet expanses of water and see a good location for pleasure boats. But boats will wash away the surface mud with their wakes and cause trees to collapse as well as frighten away the magnificent range of herons, egrets, ibises and all the other birds that make Caroni Swamp something of an ecological miracle. They see the great vistas of red mangrove trees and they envisage pathways weaving through the trees –

How long will the scarlet ibises continue to live in Trinidad's Caroni Swamp if poaching continues unchecked and the swamp is converted into a recreational park full of speed boats and other disruptions?

pathways that can only connect with other pathways, as people can hardly be expected to walk in the mud or clamber through the dense, virtually impenetrable barrier of trees. With countless creeks and tributaries that look virtually identical, visitors in boats will become completely lost unless garish signposts are erected throughout the swamp. People will fish until the fish population collapses. Huge numbers of mangrove oyster will be ripped from the red mangrove trees (roots included). It is hardly likely that the idea of converting Caroni Swamp into a recreation park is founded on a serious understanding of just how unsuitable an environment it would be for human recreation.

*Destruction of mangroves can have an international effect. Many shore birds visit specific mudflats and mangals as part of their vast migration journeys. This is true of the eastern curlew (*Numenius madagascariensis*) which is considered endangered. Thus destruction of mangroves and associated mudflats on which the species depends can only exacerbate the problems it faces.*

The destruction of a supposedly insignificant area of mangroves can impact heavily on migratory birds which use it on their migration route. Waterfowl have specific flyways and a mangrove forest and its adjoining mudflats may be suddenly and briefly home to numerous migrating birds. For example the Mai Po marshes in the New Territories of Hong Kong contain the largest remaining area of mangroves in the region; others have been destroyed for land reclamation. The Hong Kong branch of the World Wide Fund for Nature managed in 1989 to buy 90 hectares of this marsh. The shallow lagoon is a vital area for over 40 species of shorebird on their annual spring migration. Thousands visit the intertidal refuge to feed and rest.

Pollution

Oil spills can devastate mangroves. There are several ways in which oil can enter the sea as a result of human activities. The major risks are from losses

Even relatively slight tampering with the drainage characteristics of a mangal can cause the widespread death of trees. Here, in Darwin Harbour, the mangal has been disrupted for pylon (and road) building. The result is excessive salt build up due to reduction in the tidal flow: large areas of yellow mangrove have died as a result.

during the production, storage and transport of crude oil produced by the offshore oil industry. However while the media focuses on the more spectacular disasters, there are numerous other insidious ways in which oil enters the sea, and it has been estimated that about 5 million tonnes of petroleum hydrocarbons enter the seas and oceans of the world annually.

A major source of oil introduction is by tankers flushing out their tanks at sea. Although there are numerous laws to limit the amount of oil that a tanker can flush out (including processes of holding the oil in specific tanks and releasing it in sufficient dilution for it not to form a slick), the system is open to abuse.

Petroleum hydrocarbons, being organic in nature are readily broken down by bacteria. But mangroves are found in low-energy, intertidal areas. Floating oil which reaches the shore is unlikely to be broken up by wave action, so the oil coats the intertidal mud and sinks into it. Because the mud is anoxic and contains few bacteria capable of breaking down the oil it will persist for a considerable time. Furthermore, since mangroves are often in remote areas and are generally extremely difficult to work in, clean-up operations are of little effect. If the mangroves' lenticels are in oil, they will presumably become clogged and no longer be able to provide the trees with oxygen or perform the vacuum process that is vital for respiration. It has been shown that the oxygen level of mangrove roots covered in oil falls to 1 or 2 per cent of the normal concentration within two days. The animal populations living in mud saturated in petroleum hydrocarbons are also devastated. Their burrows become clogged with oil and will also provide a means for the oil to penetrate the mud to considerable depth. In other words an oil spill is likely to devastate both the animals and plants of the mangrove community.

There are many problems involved in quantifying the damage done to mangrove ecosystems by oil spills. Firstly, they often occur in remote areas

and trained observers are not readily at hand to evaluate the damage done. Soft bodied animals of the benthic infauna can die and decompose in the first few days without anyone being on hand to observe. (The black layer of oil on the substrate will also absorb heat and make the substrate too hot for numerous organisms, killing them on this account alone.)

Secondly, the resilence of a mangrove plant varies at different stages of its life. For example seedlings are nearly always recorded as being killed by oil while mature trees, although they may shed their leaves, may recover.

Thirdly, different types of oil and various elements of a single oil spill have different levels of toxicity. The aromatic components, often the most toxic, can quickly evaporate because of the high amount of solar radiation absorbed by the black coating of oil.

Fourthly, the tides, winds, currents and the geomorphology of the location will all determine the extent to which a mangal is inundated and smothered or washed clean.

*The red-collared lorikeet (*Trichoglossus rubritorquis*) of Northern Australia is a spectacularly coloured bird that forms raucous aggregations in tropical forests. Sometimes they will enter the back of the mangal to socialise. This one was babbling to a few others in a sandy mangrove tree at sunset.*

Finally the extent to which a particular mangrove forest is already stressed is likely to determine the extent to which it can survive an oil spill.

Given all these different variables it is not surprising that surveys of mangroves flooded in oil give different results. Not all are totally destroyed; some eventually recover.

Another question which results from the increased urbanisation of many estuarine areas, is what the effect is of sewage pollution on the mangrove ecosystem. Because mangroves are traditionally thought of as being worthless it is common practice in such areas to 'flush' sewage through them before it reaches the sea, a practice that has been going on in many places for many years. The sudden increase in organic input to an area provides a dramatic enrichment of nutrients, especially nitrogen and phosphorus. Other sources include run-off from agricultural fertilisers and fish-processing factories. Such organic soups will carry a heavy load of bacteria involved in the breakdown of the organic material and so the oxygen content of the water will be lowered. However the addition of nutrients can radically increase plant growth and lead to an explosion of new plants. The decomposition of the new plants, usually algae, requires a further proliferation of oxygen-hungry bacteria, and the oxygen level of the environment falls still further. A bloom in plant plankton can increase the turbidity of the system. This general process of nutrient enrichment followed by an increase in plant metabolism and a lowering of the level of available oxygen is termed eutrophication. It is a considerable problem in polluted estuaries, but it has been suggested that if sewage is flushed through mangroves they may in fact benefit from the nutrient enrichment and actually purify the sewage effluent so it reaches the sea relatively clean.

If phosphorus enters the mangal in artificially high quantities, the filtering required in the intertidal ecosystem involves the phosphorus being taken up

A sphecid wasp (Sphecidae) gathers up a ball of mud; the chambers of the nest will be stocked with captured insects or spiders.

by mangrove soils and then being utilised by the plants. Preliminary work on the chemical processes involved in the soil suggest that only small amounts of introduced phosphorus are likely to enter the substrate.

In the case of nitrogen, the mangrove ecosystem could well reduce the amount which passes through the estuary and causes eutrophication. Nitrogen entering the anoxic muds of a mangrove ecosystem is usually incorporated as the ammonium ion: the lack of oxygen prevents further oxidation to nitrate. However the ammonium ion can diffuse out of the anaerobic layer of soil into the aerobic layer where aerobic bacteria oxidise it to nitrate. This process is termed nitrification. The nitrate can now diffuse back into the anaerobic layer where it is reduced by chemical and anaerobic bacterial action to gaseous nitrogen or nitrous oxide. This then diffuses out of the soil into the water column. The process, which is discussed in chapter 2, is termed denitrification and means that the nitrogen component of the soil is steadily lost to the surroundings. It appears

*The brown honeyeater (*Lichmera indistincta*) is one of the more common honeyeaters in Northern Territory mangals. The female builds her nest just above the high tide level and lines it with silk from spider webs.*

that the addition of higher levels of nitrogen from sewage or fertiliser actually increases the rate of denitrification. It may therefore be that the usable nitrogen component of the effluent reaching an estuary is effectively reduced. However there are numerous variables, such as the acidity of the soil in question, and there is a possibility that the chemical processes involved in the mangrove soils change their constitution, making them increasingly less efficient at denitrification.

Will mangroves benefit from higher levels of nitrogen and phosphorus derived from effluent? It will be remembered that trees regularly used by nesting or roosting birds and therefore subjected to a high level of guano appear to grow faster and flower earlier than their less fortunate neighbours. Similarly one might expect mangrove trees subjected to nitrogen- and phosphorus-rich sewage to benefit – or at least not to be harmed. In Darwin, sewage is treated in three oxidation ponds before being discharged into the mangroves. Work done there in the 1980s suggested that trees in the path of the

*An immature yellow-crowned night heron (*Nycticorax violaceus*) hunts for food on a mangrove-fringed shoreline in Bimini, the Bahamas. As this photograph shows, not all night herons are nocturnal!*

An upside-down jellyfish (Cassiopeia xamachana) floats past a red mangrove tree in Bimini, Bahamas. These jellyfishes often line the floor of tidal creeks just below the low water mark. They lie, as the common name suggests, in an upside down position with the algae-filled tentacles pointing towards the surface. The jellyfish benefits from the photosynthetic activity of its algae and the algae have a relatively safe home.

sewage were taller around the effluent outfalls, although this could also be due to the fact that they receive a regular supply of fresh water from the sewage system. Furthermore the small-leafed orange mangrove was observed to be more common (and taller) in the tidal creek bank areas influenced by the sewage outfalls, which might suggest that sewage can also affect the species composition of the mangal. Surveys done by Russell Hanley on the animals of this area suggested that their populations were also healthy. In other words the sewage outfalls were having no ill effect in this area.

The above example suggests that the increased availability of nitrogen and phosphorus within the mangal can benefit the trees, although it is possible that the trees may not be removing sufficient amounts to reduce the risk of eutrophication in the estuary significantly. This does not mean, however, that mangroves offer a miracle filtration system for untreated sewage. The high bacterial content of the effluent, which feeds on the inorganic carbon, can radically reduce the oxygen content of the water, which could suffocate the aquatic animal component of the mangal. Furthermore the increased supply of organic carbon entering the soil can increase the anaerobic microbial activity and alter the chemical properties of the soil as a result, making it an even more reductive medium. Mangrove plants grow within a range of anaerobic soil conditions and it is probable that a soil burdened with excessive organic carbon could be transformed into one inimical to plant growth.

This white-bellied cuckoo-shrike (Coracina papuensis) *has captured a grasshopper in a Darwin Harbour mangrove.*

Heavy metals such as tin, lead and mercury are not usually required for the metabolic processes of organisms. Nevertheless they are taken in by both plants and animals and can be highly toxic in low concentrations. They can enter the sea by the discharge of industrial effluent. Anaerobic soils tend to trap heavy metals in a variety of ways including by reactions with the sulphur component of the soil and the precipitation of insoluble sulphides. So mangrove soils might act as heavy metal sinks. However it would be grossly irresponsible to assume that industrial effluent could therefore simply be pumped into intertidal forests, as there are still likely to be processes whereby the heavy metals can enter the organisms of the mangal with disastrous effects.

Anaerobic soils will adsorb (attract to their surfaces) and denature pesticides such as DDT and lindane. Thus they may act as a buffer for pesticide discharge. The effectiveness of such a buffer would depend in part on the area of anaerobic soil to which the pesticide is exposed before it leaves that soil type as well as the concentration of the pesticide. This in turn would depend on tides, currents and the

Cook's tree boa (Corallus enhydris) *hunts small birds in the mangroves such as the bicoloured conebill* (Conirostrum bicolor) *as well as small mammals. Caroni Swamp.*

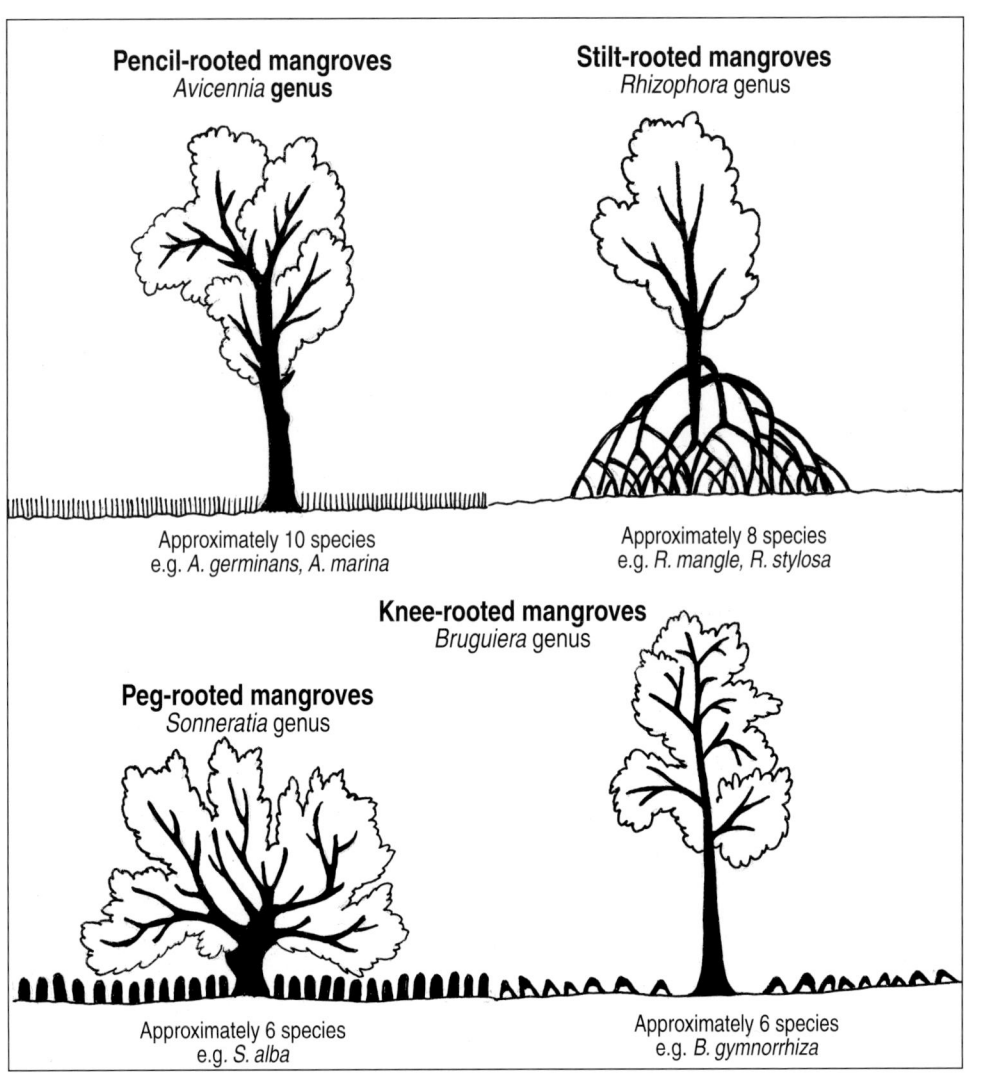

area, depth and permeability of the anaerobic soil available. The extent to which anaerobic soils can be used to degrade pesticides is not yet clear; and their effect on the mangal community before they are denatured is yet to be ascertained.

The scale of the problem

The pressures facing the mangrove ecosystems of the world are many and varied. The clear felling of vast tracts of mangrove is well documented. However many mangals have suffered a less conspicuous decline as a result of human intrusions. The rate of loss is deeply alarming. A report prepared by IUCN (The World Conservation Union) and UNEP (United Nations Environment Programme) in 1985 estimated that 50 per cent of the mangroves of the Indian Ocean had been destroyed in the previous decade alone.

In the Philippines there were estimated to be as much as 500,000 hectares of mangroves in 1918. By 1987 the figure was reduced to about 100,500 hectares. In Thailand mangroves have been reduced from 367,000 hectares in 1961 to 287,300 by 1979. In 1963 some 682,000 hectares of mangrove fringed the Indian sub-continent. By 1977 only 365,500 hectares remained. At the height of the destruction of Indonesian mangroves for wood chips an area of 2000 sq km of mangrove was being lost annually.

The mangrove *Heritiera fomes* forms huge stands in the largest mangrove forest of the world: the Sundarbans of the Ganges Delta. It requires brackish conditions in order to survive. However the building of the Farraka Barrage on the Ganges has caused a thirteen-fold increase in salinity downstream which has devastated large areas of the Sundarban mangal.

The figures for New World mangroves are hardly any better: 75 per cent of Puerto Rican mangroves are thought to have been destroyed. In the 20 years from 1960 to 1980 over a quarter of Ecuador's mangroves were destroyed for

Half of the world's mangroves have been destroyed in the last 50 years. What do the next 50 years hold in store?

shrimp ponds. On a global scale it has been suggested that between 40 and 50 per cent of the world's mangroves have been destroyed in the last 50 years.

Are there solutions?

If the destruction of mangroves made economic sense – if it were possible to argue that a couple of years' worth of wood chips or charcoal or farmed prawns was more valuable than the long term future supply of offshore catches - then at least the destroyers would have some sort of an argument. It would not be a very good argument, but it would be a justification of sorts. But we have seen time and time again how the destruction of mangroves is economic as well as environmental lunacy.

It has been estimated that mangroves exploited at a subsistence level can support ten times as many people as they can when exploited commercially. And this estimate is probably on the conservative side in so far as subsistence exploitation presupposes sustainablity and can endure for generations whereas commerical exploitation, like the crudest of parasites, tends to destroy the very thing on which it depends. And when developers destroy mangrove forests for a quick profit their calculations do not include the value of low-lying lost land because of the subsequent invasion of the sea, nor the cost of building and maintaining artificial barriers to protect the shoreline from accelerated erosion - a job the mangroves did naturally.

There are, however, some positive signs. Planting schemes are afoot to try to re-establish mangroves by planting their seedlings. In those areas where monospecific stands were the norm the chances of some sort of recovery appear fair. However since we do not know the processes that cause zonation, the chances of being able to replicate zonation artificially do not look very good. And zonation is crucial if the faunal component is to recover to its original level.

So what can be done? The general public must be better educated about the value of mangroves; in many parts of the world this is already being done. While access in slow-moving boats should be encouraged, large numbers of people should not be allowed to walk in the mangroves as it is impossible to avoid damage to the root systems of the plants and the crushing of numerous organisms in the mud. Specific areas of mangal should be set aside for educational purposes and walkways constructed sensitively to minimise damage to the trees, so that people can gain a first-hand appreciation of this ecosystem in comfort and safety. Tropical areas which derive substantial income from tourism would then, for relatively little cost, add a vital educational element to the tourists' visit. Walkways could lead to hides facing bird-rich intertidal mudflats so that bird-lovers could observe the waterbirds without disturbing them.

Given that half of the world's mangroves have been destroyed to no lasting good, any rational scheme for their management must be primarily concerned with conserving rather than exploiting them. In other words the onus should be on developers to justify the need for development. This would require something that has been singularly lacking to date: an adequate evaluation of the worth of this ecosystem in terms of the multiplicity of benefits it provides not only to humans (at a subsistence level) but to other organisms and indeed to other ecosystems as well.

If there is a single feature of the magnificent intertidal kingdom of mangroves that I will remember longest it is not the other-worldly trees nor the dancing creatures of the mud vanishing in panic at my clumsy approach, nor the outraged shriek of fleeing birds, nor even the sloppy scurrying of a terrifying crocodile. Rather it is the silence haunted by my intrusion; a silence I have violated like a poltergeist; a silence that asks a simple question of all who enter prepared to listen: friend or foe?

Bibliography

Allen, G.R. and Swainston, R. (1988). *The Marine Fishes of North-Western Australia*. Western Australian Museum, Perth.

Andrews, T.J., Clough, B.F. and Muller, G.J. (1984). 'Photosynthetic gas exchange properties and carbon isotope ratios of some mangroves in North Queensland' in *Physiology and Management of Mangroves*, ed. Teas, H.J., Tasks for vegetation science No. 9, Dr W. Junk, The Hague.

Barnes, R.S.K. and Hughes, R.N. (1988) *An Introduction to Marine Ecology*. Blackwell Scientific Publications, Oxford.

Barnwell, F.H. (1968.) 'The Role of Rhythmic Systems in the Adaption of Fiddler Crabs to the Intertidal Zone', Am. Zoologist, 8: 569–583.

Bhosale, L.J. and Shinde, L.S. (1983). 'Significance of cryptovivipary in *Aegiceras corniculatum* (L.) Blanco' in *Biology and Ecology of Mangroves* ed. Teas, H.J., Tasks for vegetation science No. 8, Dr W. Junk, The Hague.

Bird, E.C.F., (1986). 'Human Interactions with Australian Mangrove Ecosystems' in *Man in the Mangroves,* ed. Kunstadter P. et al. The United Nations University, Tokyo.

Blasco, F. (1984). 'Climatic factors and the biology of mangrove plants' in *The Mangrove Ecosystem: Research Methods*, ed. Snedaker, S.C. and Snedaker, J.G. Unesco, Paris.

Blasco, F. (1984). Mangrove evolution and palynology' in *The Mangrove Ecosystem: Research Methods*, ed. Snedaker, S.C. and Snedaker, J.G. Unesco, Paris.

Blasco, F. (1984). 'Taxonomic considerations of the mangrove species' in *The Mangrove Ecosystem: Research Methods*, ed Snedaker, S.C. and Snedaker, J.G. Unesco, Paris.

Boaden, P.J.S. and Seed, R. (1985). *An Introduction to Coastal Ecology*. Blackie & Son, Glasgow.

Bohlke, J.E. and Chaplin C.C.G. (1970). *Fishes of the Bahamas*. Livingston, Wynnewood.

Bond, J. (1985). *Birds of the West Indies*. Collins, London.

Borror, D.J. and White, R.E. (1970). *A Field Guide to Insects*. Petersen Field Guides Vol. 19, Houghton Mifflin, Boston.

Boto, K.G. (1984). 'Waterlogged saline soils' in *The Mangrove Ecosystem: Research Methods*, ed. Snedaker, S.C. and Snedaker, J.G. Unesco, Paris.

Brown, M.S. (1984). 'Mangrove leaf litter production and dynamics' in *The Mangrove Ecosystem: Research Methods*, ed. Snedaker, S.C. and Snedaker, J.G. Unesco, Paris.

Chan, H.T. (1986). 'Human Habitation and Traditional Uses of the Mangrove Ecosystem in Peninsular Malaysia' in *Man in the Mangroves,* ed. Kunstadter P. et al. The United Nations University, Tokyo.

Chapman, V.J. (1976). *Coastal Vegetation*. Pergamon Press, Oxford.

Chapman, V.J. (1984). 'Botanical surveys in mangrove communities' in *The Mangrove Ecosystem: Research Methods*, ed. Snedaker, S.C. and Snedaker, JG. Unesco, Paris.

Cintron, G. and Novelli, Y.S. (1984). 'Methods for studying mangrove structure' in *The Mangrove Ecosystem: Research Methods*, ed. Snedaker, S.C. and Snedaker, J.G. Unesco, Paris.

Clark, R.B. (1986). *Marine Pollution*. Clarendon Press, Oxford.

Clough, B.F., Boto, K.G. and Attiwill, P.M. (1983). 'Mangroves and sewage: a re-evaluation' in *Biology and Ecology of Mangroves*, ed. Teas, H.J. Tasks for vegetation science No. 8, Dr W. Junk, The Hague.

Collette, B.B. (1983). 'Mangrove fishes of New Guinea' in *Biology and Ecology of Mangroves*, ed. Teas, H.J., Tasks for vegetation science No 8, Dr W. Junk, The Hague.

Conant, R. and Collins, J.T. (1991). *Reptiles and Amphibians of Eastern/ Central North America*. Petersen Field Guides Vol. 12, Houghton Mifflin, Boston.

Cooper, St.G.C. and Bacon, P.R. (eds.) (1981). *The Natural Resources of Trinidad and Tobago*. Edward Arnold, London.

Cronin, L. (1989). *Key Guide to Australian Palms Ferns and Allies*. Reed, Frenchs Forest.

Day, J.W. Jr et al. (1989). *Estuarine Ecology*. Wiley, New York.

DePatra K.D. and Levin, L.A. (1989). 'Evidence of the passive deposition of meiofauna into fiddler crab burrows'. *J.Exp. Mar. Biol. Ecol.*, 125: 173–192.

Dwivedi, S.N. and Padmakumar, K.G. (1983). 'Ecology of a mangrove swamp near Juhu Beach, Bombay with reference to sewage pollution' in *Biology and Ecology of Mangroves*, ed. Teas, H.J. Tasks for vegetation science No 8, Dr W. Junk, The Hague.

Elsol, J.A. and Saenger, P. (1983). 'A general account of the mangroves of Princess Charlotte Bay with particular reference to zonation of the open shoreline' in *Biology and Ecology of Mangroves*, ed. Teas, H.J. Tasks for vegetation science No 8, Dr W. Junk, The Hague.

Emmons, L.H. (1990). *Neotropical Rainforest Mammals*. The University of Chicago Press, Chicago.

Farnworth, E.G. and Golley, F.B. (eds.) (1973). *Fragile Ecosystems: Evaluation of Research and Applications in the Neotropics*. A report of the Institute of Ecology (TIE), Springer-Verlag, Berlin.

Fell, J.W., Master, I.M. and Wiegert, R.G. (1984). 'Litter decomposition and nutrient enrichment' in *The Mangrove Ecosystem: Research Methods*, ed. Snedaker, S.C. and Snedaker, J.G. Unesco, Paris.

Field, C.D., Hinwood, B.G. and Stevenson, I. (1984). 'Structural features of the salt gland of *Aegiceras*' in *Physiology and Management of Mangroves*, ed. Teas, H.J. Tasks for vegetation science No 9, Dr W. Junk, The Hague.

ffrench, R. (1992). *A Guide to the Birds of Trinidad and Tobago*. Christopher Helm, London.

ffrench, R. (1992). *Birds of Trinidad and Tobago*. Macmillan, London.

Gearing, J.N. (1988). 'The use of stable isotope ratios for tracing the nearshore-offshore exchange of organic matter' in *Lecture Notes on Coastal and Estuarine Studies* Vol. 22, ed. Jansson, B-O. Springer-Verlag, Berlin.

George, R.W. and Jones, D.S. (1982). 'A revision of the fiddler crabs of Australia (Ocypodinae: *Uca*)'. *Records of the Western Australian Museum Supplement* No. 14.

Goldsmith, E., Hilyard, N., Bunyard, P. and McCully, P. (1990). *5000 Days to Save the Planet*. Hamlyn, London.

Goode, J. (1980). *Insects of Australia*. Angus and Robertson, London.

Gomez, P.L. (1988). 'Production and transport of organic matter in mangrove-dominated estuaries' in *Lecture Notes on Coastal and Estuarine Studies*, Vol. 22, ed.Jansson, B-O. Springer-Verlag, Berlin.

Hagen, H-O. von and Jones, D.S. (1989). 'The fiddler crabs (Ocypodidae: *Uca*) of Darwin, Northern Territory, Australia' in *The Beagle, Records of the Northern Territory Museum of Arts and Sciences*, 6 (1): 55–68.

Hagen, H-O. von (1993) 'Waving display of *Uca polita* and other Australian fiddler crabs'. *Ethology* 93, 3-20.

Hancock, J. and Kushlan, J. (1984). *The Herons Handbook*. Croom Helm, London.

Hayman, P., Marchant, J. and Prater T. (1991). Shorebirds *An Identification Guide to the Waders of the World* . Christopher Helm, London.

Holthuis, L.B. (1991). *Marine Lobsters of the World*. FAO species catalogue Vol. 13, Food and Agriculture Organisation of the United Nations, Rome.

Hutchings, P.A. and Recher, H.F. (1983). 'The faunal communities of Australian mangroves' in *Biology and Ecology of Mangroves*, ed. Teas, H.J., Tasks for vegetation science No. 8, Dr W. Junk, The Hague.

Hutchins, P. and Saenger, P. (1987). *Ecology of Mangroves*. University of Queensland Press, Townsville.

Irby, B.N. et al (1984). *Diversity of Marine Animals* in Man and the Gulf of Mexico Series, Mississippi-Alabama Sea Grant Consortium, University of Mississippi, Jackson.

IUCN/UNEP (1985) *Management and Conservation of Renewable Marine Resources in the Indian Ocean Region; overview*. (Written and edited by Tropical Marine Research Unit, Dept. of Biology, University of York.) UNEP Regional Seas reports and studies No. 60.

Janson, B-O. (ed.) (1988). 'Coastal-offshore interactions - an evaluation of presented evidence' in *Lecture Notes on Coastal and Estuarine Studies* Vol. 22. Springer-Verlag, Berlin.

Jernelov, A. and Linden, O. (1983). 'The effects of oil pollution on mangroves and fisheries in Ecuador and Columbia' in *Biology and Ecology of Mangroves*, ed. Teas, H.J. Tasks for vegetation science No. 8, Dr W. Junk, The Hague.

Johnstone, I.M. (1983). 'Succession in zoned mangrove communities: where is the climax?' in *Biology and Ecology of Mangroves*, ed. Teas, H.J. Tasks for vegetation science No. 8, Dr W. Junk, The Hague.

Johnstone, R.E. (1990). 'Mangroves and mangrove birds of Western Australia'. *Records of the Western Australian Museum*, Supplement No. 32, Perth.

Joshi, G.V. et al (1984). 'Photosynthesis and photorespiration in mangroves' in *Physiology and Management of Mangroves*, ed. Teas, H.J, Tasks for vegetation science No. 9, Dr W. Junk, The Hague.

Kaplan, E. H. (1988). *A Field Guide to Southeastern and Caribbean Seashores*. Petersen Field Guides Vol. 36, Houghton Mifflin, Boston.

Kunstadter, P. et al (ed.) (1986). 'Socio-economic and demographic aspects of mangrove settlement' in *Man in the Mangroves*. The United Nations University, Tokyo.

Lane, B. and Davies, J. (1987). *Shorebirds in Australia*. Nelson, Melbourne.

Lear, R. and Turner, T. (1977). *Mangroves of Australia*. University of Queensland Press.

Lehninger, A.L. (1976). *Biochemistry*. Worth, New York.

Lewis, R.R. (1983). 'Impact of oil spills on mangrove forests' in *Biology and Ecology of Mangroves*, ed. Teas, H.J. Tasks for vegetation science No. 8, Dr W. Junk, The Hague.

Lincoln, R.J. and Boxshall, G.A. (1990). *The Cambridge Illustrated Dictionary of Natural History*. Cambridge University Press, Cambridge.

Lowe-McConnell, R.H. (1987). *Ecological Studies in Tropical Fish Communities.* Cambridge University Press, Cambridge.

Lovelock, C. (1993). *Field Guide to the Mangroves of Queensland.* Australian Institute of Marine Science.

Macintosh, D.J. (1979). 'Predation of fiddler crabs [*Uca* spp.] in estuarine mangroves' in *Mangrove & Estuarine Vegetation in Southeast Asia.* Biotropical Special Publication No. 10.

Macnae, W. (1968). 'Fauna and flora of mangrove swamps'. *Advances in Marine Biology*, Vol. 6, Academic Press.

Marshall, A.J. and Williams, W.D. (1975). *Textbook of Zoology Invertebrates.* Macmillan Press, London.

Mascord, R. (1991). *Australian Spiders in Colour.* Reed, Balgowlah. NSW.

Mattison, C. (1992). *Snakes of the World.* Blandford, London.

McGuiness, K.A. (1990). 'Effects of oil spills on macro-invertebrates of saltmarshes and mangrove forests in Botany Bay, New South Wales, Australia' in *J. Exp. Mar. Biol. Ecol.*, 142: 121–135.

Miller, D.C. (1961). 'The feeding mechanism of fiddler crabs, with ecological considerations of feeding adaptions'. *Zoologica*, 46:8. New York Zoological Society.

Mirtschin, P. and Davis, R. (1992). *Snakes of Australia.* Hill of Content, Melbourne.

Murray, P.F. and Hanley, J.R. (1986). 'Unmuddling the mudlobster; observations on the age and taphonomy of fossil *Thalassina*' in *The Beagle, Occasional Papers of the Northern Territory Museum of Arts and Sciences*, 3 (1): 59–70, Darwin.

Museums and Art Galleries of the Northern Territory. *Fossil Mudlobsters.* Information Leaflet No. 6, Museums and Art Galleries of the Northern Territory, Darwin.

Museums and Art Galleries of the Northern Territory. *Mudskippers.* Information Leaflet No. 18, Museums and Art Galleries of the Northern Territory, Darwin.

Norman, J.R. (1975). *A History of Fishes.*, 3rd edition. Revised by P.H. Greenwood, Ernest Benn, London.

Nursall, J.R. (1981). 'Behaviour and habitat affecting the distribution of five species of sympatric mudskippers in Queensland'. *Bulletin of Marine Science*, Vol. 31 (3).

Palmer, J.D. (1988). Comparative studies of tidal rhythms. VI. Several clocks govern the activity of two species of fiddler crabs'. *Marine Behavioural Physiology*, 13: 201–219.

Pannier, F. (1984). 'Mangrove physiology: photosynthesis' in *The Mangrove Ecosystem: Research Methods*, ed. Snedaker, S.C. and Snedaker, J.G. Unesco, Paris.

Pannier, F. (1984). 'Mangrove physiology: water relations' in *The Mangrove Ecosystem: Research Methods*, ed. Snedaker, S.C. and Snedaker, J.G. Unesco, Paris.

Percival, M. and Womersley, J.S. (1975). Floristics and ecology of the mangrove vegetation of Papua New Guinea'. *Botany Bulletin* No. 8, Department of Forests, Division of Botany, Lae.

Pizzey, G. (1988.) *A Field Guide to the Birds of Australia*. Collins, Sydney.

Polunin, I. (1972). 'Who says fish can't climb trees?' *National Geographic Magazine*, 141 (1) The National Geographic Society, Washington.

Preston-Mafham, R. (1991). *Spiders An Illustrated Guide*. Quarto, London.

Raymond, A. and Phillips, T.L. (1983.) 'Evidence for an Upper Carboniferous mangrove community' in *Biology and Ecology of Mangroves*, ed. Teas, H.J. Tasks for vegetation science No. 8, Dr W. Junk, The Hague.

Reader's Digest (1984). *Reader's Digest Book of the Great Barrier Reef*. Reader's Digest, Sydney.

Reid, D.G. (1986). *The Littorinid Molluscs of Mangrove Forests in the Indo-Pacific Region*. British Museum (Natural History), London.

Reynolds, J.E. and Odell, D.K. (1991). *Manatees and Dugongs*. Facts on File, New York.

Ross, C.A. and Garnett, S. et al. *Crocodiles and Alligators*. Merehurst Press, London.

Schoener, T.W. (1968). 'The Anolis lizards of Bimini: resource partitioning in a complex fauna'. *Ecology*, 49 (4).

Sheppard, C., Price, A. and Roberts, C. (1992). *Marine Ecology of the Arabian Region*. Academic Press, London.

Simpson K. and Day, N. (1991). *Field Guide to the Birds of Australia*. Christopher Helm, London.

Smith, T.J. et al (1991). 'Keystone species and mangrove forest dynamics: the influence of burrowing by crabs on soil nutrient status and forest productivity' in *Estuarine, Coastal and Shelf Science*, 33: 419–432. Australian Institute of Marine Science, Townsville.

Snedaker, J. (1987). *Mangrove Mythology*. Florida Naturalist.

Snedaker, S.C. (1986). 'Traditional uses of South American mangrove resources and the socio-economic effect of ecosystem changes' in *Man in the Mangroves*, ed. Kunstadter P. et al. The United Nations University, Tokyo.

Steinke, T.D. and Charles, L.M. (1984). 'Productivity and phenology of *Avicennia marina* (Forsk.) Vierh. and *Bruguiera gymnorrhiza* (L.) Lam. in Mgeni estuary, South Africa in *Physiology and Management of Mangroves*, ed. Teas, H.J. Tasks for vegetation science No. 9, Dr. W. Junk, The Hague.

Steinke, T.D., Naidoo, G. and Charles, L.M. (1983). 'Degradation of mangrove leaf and stem tissues in situ in Mgeni Estuary, South Africa' in *Biology and Ecology of Mangroves*, ed. Teas, H.J. Tasks for vegetation science No. 8, Dr W. Junk, The Hague.

Stirling, P.D. (1986). *Butterflies and Other Insects of the Eastern Caribbean*. Macmillan, London.

Tait, R.V. (1983). *Elements of Marine Ecology*. Butterworths, London.

Thom, B.G. (1984.) 'Coastal landforms and geomorphic processes' in *The Mangrove Ecosystem: Research Methods*, ed. Snedaker, S.C. and Snedaker, J.G. Unesco, Paris.

Tomlinson, P.B. (1986). *The Botany of Mangroves*. Cambridge University Press, Cambridge.

Twilley, R. (1988). 'Coupling of mangroves to the productivity of estuarine and coastal waters' in *Lecture Notes on Coastal and Estuarine Studies*, Vol. 22, Jansson, B-O (ed.), Springer-Verlag, Berlin.

Warren, J.H. (1990). 'Role of burrows as refuges from subtidal predators of temperate mangrove crabs'. *Marine Ecology Progress Series*, 67: 295–299.

Wells, A.G. (1983). 'Distribution of mangrove species in Australia' in *Biology and Ecology of Mangroves*, ed. Teas, H.J. Tasks for vegetation science No. 8, Dr W. Junk, The Hague.

Wenban-Smith, M.G. (1993). 'Global status and extent of mangrove forests'. Report commissioned by WWF-UK.

Wightman, G.M. (1989). 'Mangroves of the Northern Territory'. *Northern Territory Botanical Bulletin*, 7. Conservation Commission of the Northern Territory, Palmerston.

Williams, D.D. and Feltmate, B.W. (1992). *Aquatic Insects*. CAB International, Wallingford.

Woodroffe, C.D. (1983.) 'Development of mangrove forests from a geological perspective' in *Biology and Ecology of Mangroves*, ed. Teas, H.J. Tasks for vegetation science No. 8, Dr W. Junk, The Hague.

Womersley, J.S. (1983). 'An introduction to the nomenclature and taxonomy of the mangrove flora in Papua New Guinea and adjacent areas' in *Biology and Ecology of Mangroves*, ed. Teas, H.J. Tasks for vegetation science No. 8, Dr W. Junk, The Hague.

Womersley, J.S. (1984). 'Observations on water salinity in mangrove associations at two localities in Papua New Guinea' in *Physiology and Management of Mangroves*, ed. Teas, H.J. Tasks for vegetation science No. 9, Dr W. Junk, The Hague.

![I]ndex